„Advances in Polymer Science/Fortschritte der Hochpolymeren-Forschung"

erscheinen zwanglos in einzeln berechneten Heften, die zu Bänden vereinigt werden.

Sie enthalten Fortschrittsberichte monographischen Charakters aus dem Gebiet der Physik und Chemie der Hochpolymeren mit ausführlichen Literaturzusammenstellungen. Sie sollen der Unterrichtung der auf diesen Gebieten Tätigen über solche Themen dienen, die in letzter Zeit besondere Aktualität gewonnen haben, bzw. die in neuerer Zeit eine lebhafte und nach literarischer Zusammenfassung verlangende Entwicklung erfahren haben.

Anschriften der Herausgeber:

Prof. Dr. J. D. Ferry, Department of Chemistry, The University of Wisconsin, Madison 6, Wisconsin/USA.

Prof. Dr. W. Kern, Institut für Organische Chemie der Universität, Mainz.

Prof. Dr. G. Natta, Istituto di Chimica Industriale del Politecnico, Milano/Italien.

Prof. Dr. C. G. Overberger, Polytechnic Institute of Brooklyn, 333 Jay Street, Brooklyn 1, New York/USA.

Prof. Dr. G. V. Schulz, Institut für physikalische Chemie der Universität, Mainz.

Prof. Dr. A. J. Staverman, Fruinlaan 6, Leiden/Holland.

Prof. Dr. H. A. Stuart, Institut für physikalische Chemie der Universität, Mainz.

Springer-Verlag

69 Heidelberg 1	1 Berlin 31 (Wilmersdorf)
Postfach 3027	Heidelberger Platz 3
Fernsprecher 4 91 01	Fernsprecher 83 03 01
Fernschreiber 04-61 723	Fernschreiber 01-83 319

---

4. Band · **Inhaltsverzeichnis** · 1. Heft

ISBN 978-3-540-03245-8   ISBN 978-3-540-37126-7 (eBook)
DOI 10.1007/978-3-540-37126-7

Adv. Polymer Sci., Vol. 4, pp. 1—65 (1965)

The Kinetics and Mechanism of N-carboxy-α-amino-acid Anhydride (NCA) Polymerisation to Poly-amino Acids

By

M. Szwarc[1]

Donnan Chemical Laboratories
The University, Liverpool, England

With 22 Figures

Table of Contents

1. Introduction

In reviewing the polymerisation of NCA's the writer has had three aims in mind. Firstly, he has tried to acquaint the reader with the factual material and then to deduce from it the most satisfactory mechanism for these polymerisations. However, he wanted also to show how the various approaches to these processes developed, to survey the reported hypotheses — why they were postulated, how they were tested and perhaps rejected. He believes that one learns more from failures than from achievements and therefore he has not hesitated to review at length even those ideas which eventually have been proved unsuitable for the systems for which initially they were conceived. However, an interesting idea, although unattainable in its original domain, may turn out to be profitable in another and hence, its review may be desirable, since it might inspire other investigators and suggest to them useful approaches

[1] Permanent address: Dept. of Chemistry, New York State, College of Forestry, Syracuse, N.Y./U.S.A.

to their problems. To achieve these goals it has been necessary to maintain a somewhat historical attitude which may reveal more clearly the thoughts leading to the proposed hypotheses.

Polymerisation of N-carboxy-α-amino acid anhydrides to polypeptides represents an important and interesting process which yields polymers not easily produced by other techniques. This reaction, which has attracted the attention of many investigators and has been extensively studied during the last 20 years, proceeds through a chain polyaddition, i. e. the growth is determined by the sequence of steps $P_n + M \rightarrow$ $\rightarrow P_{n+1}$ rather than $P_n + P_m \rightarrow P_{n+m}$ (see however p. 39). The overall reaction proceeds through initiation and propagation steps and often it includes some termination.

N-carboxy-α-amino-acid anhydrides react in a variety of ways and, therefore, it is difficult to discuss their polymerisation in a conventional manner, i. e. by separately considering the initiation, propagation and termination steps. It is believed that the reader more easily may comprehend its nature if the material is presented in terms of the chemical behaviour of the monomer towards various reagents. This course, therefore, will be followed in the subsequent sections.

2. Leuchs' anhydrides (NCA)

The N-carboxy-α-amino-acid anhydrides, referred to as Leuchs' anhydrides, or briefly as NCA's, are well-defined, colourless, crystalline substances having sharp melting points when pure. The synthesis of the simplest member of this class of compounds, viz. oxazolidine-2,5-dione,

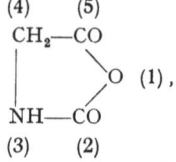

was described by Leuchs (1) in 1906 and in subsequent papers (2) he reported the preparation of some of its derivatives, namely the N-carboxy anhydrides of N-phenyl glycine, C-phenyl glycine, phenyl alanine and leucine. Numerous derivatives of oxazolidine-2,5-dione have been synthesised since, and an impressive list of about 100 Leuchs' anhydrides, together with references to their preparation and melting points, is given in a recent review by Katchalski and Sela (3).

The Leuchs' anhydrides are extremely reactive compounds, and most of them may be stored in a dry state only below −20 or −30° C. At 10° C they dissolve in water with little decomposition[1], but eventually

---

[1] This reaction is greatly affected by pH of the solution. For example, an extensive polymerisation occurs at pH = 7.4 (see ref. 63).

they hydrolyse into carbamic acids,

$$\begin{array}{ccc} R \cdot CH{-}CO & & R \cdot CH \cdot COOH \\ \big| \quad \diagdown\!\!O + H_2O \rightarrow & & \big| \\ NH{-}CO & & NH \cdot COOH, \end{array}$$

which readily decarboxylate into the respective amino-acids. The kinetics of the hydrolysis and decarboxylation of NCA's in aqueous solution was investigated by BARTLETT and JONES (4). The reaction shows a first order character, the respective rate constants being 10^{-3} — 10^{-2} sec^{-1} at $25°$ C. However, acids and bases catalyse this process.

On exposure to moisture, the crystalline anhydrides undergo an auto-accelerated polymerisation with evolution of carbon dioxide. This solid-state reaction proceeds without any observable change in the form of the crystal, and its effect on the X-ray pattern has been briefly investigated by MILLER, FANKUCHEN and MARK (5). The probable mechanism of this process will be discussed later.

On melting the crystals, a rapid decarboxylation takes place with the formation of a low molecular weight polypeptide. These spontaneous polymerisations were first noticed by LEUCHS (1, 2), and were subsequently thoroughly studied by KATCHALSKI and his co-workers (6) who carried out these reactions in vacuum at elevated temperatures. Under these conditions the conversion of the anhydride into polypeptide appears to be quantitative.

The mechanism of high temperature bulk polymerisation of NCA's is still obscure. It was suggested that the reaction involves the isomerisation of the anhydride into isocyanate, viz.

$$\begin{array}{ccc} R \cdot CH \cdot CO & & R \cdot CH \cdot COOH \\ \big| \quad \diagdown\!\!O \;\leftrightharpoons & & \big| \\ NH \cdot CO & & N{=}C{=}O, \end{array}$$

since the reverse step takes place in the Curtius' synthesis of NCA. The polymerisation of isocyanates to linear anhydrides,

$$\begin{array}{ccccc} HO \cdot CO \cdot CH \cdot R & & N{=}C{=}O & HO \cdot CO \cdot CH \cdot R & N{=}C{=}O \\ \big| & + & \big| & \rightarrow \quad \big| & \big| \\ N{=}C{=}O & & HO \cdot CO \cdot CH \cdot R & NH \cdot CO \cdot O \cdot CO \cdot CH \cdot R \end{array}$$

may be facile[1] and the subsequent rapid decarboxylation of the latter is known to give a peptide bond, i. e.

$$HO \cdot CO \cdot CH(R) \cdot NH \cdot CO \cdot O \cdot CO \cdot CH(R) \cdot N{=}C{=}O \rightarrow$$
$$\rightarrow HO \cdot CO \cdot CH(R) \cdot NH \cdot CO \cdot CH(R) \cdot N{=}C{=}O + CO_2 .$$

These decarboxylations were extensively studied by DICKMANN and BREEST (7) who elucidated their mechanism. The terminal isocyanate

---

[1] The reaction $R \cdot N{=}C{=}O + HO \cdot CO \cdot R_1 \rightarrow R \cdot NH \cdot CO \cdot O \cdot CO \cdot R_1$ has to be catalysed by bases. It is not known whether the presence of adjacent carboxylic groups would facilitate this process.

group of the resulting polymer may be easily lost, e. g., by its reaction with moisture, and therefore it is not surprising that its presence was undetected in the final product. The isocyanate mechanism resembles that proposed once by Kopple (*8*) for the solution polymerisation of NCA.

The thermal decarboxylation of NCA's involves the 2-carbonyl group of the oxazolidine-2,5-dione ring. This was proved (*9*) by labelling glycine NCA with C^{13} and examining the C^{13} content of the carbon dioxide evolved in thermal polymerisation.

The reaction of Leuchs' anhydrides with primary and secondary amines proceeds in two ways, viz.

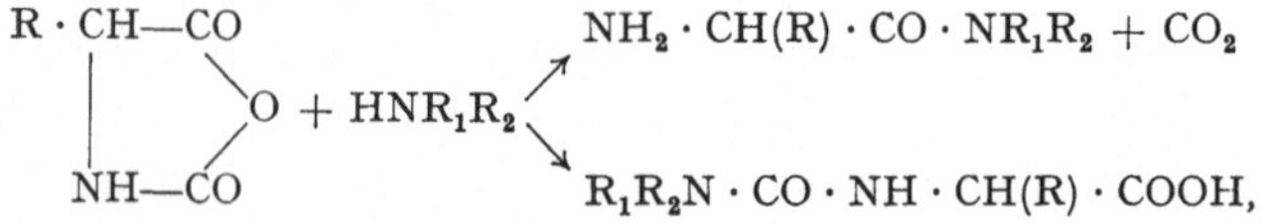

(a)

and the detailed mechanism of these additions is the subject of the following section.

3. Reactions of Leuchs' anhydrides with primary and secondary amines

The reaction of NCA's with amines was reported by Fuchs (*10*) in 1922 and by Wessely (*11*) in 1925. From the resulting mixture they isolated amides formed by the interaction such as

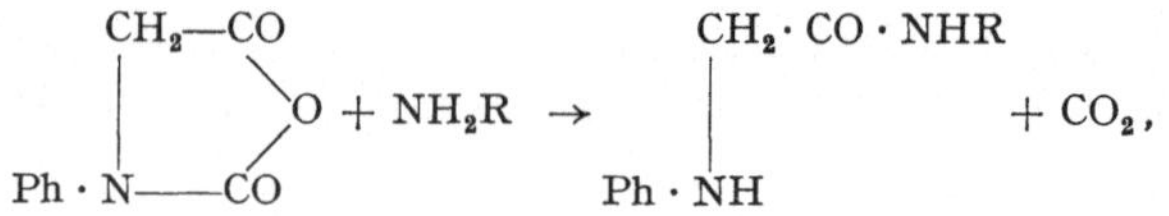

and, as shown by the subsequent work, these substances are the main products of the addition if the amine is in excess. For example, 90% of phenyl alanine amide was obtained when phenyl alanine NCA was reacted with an excess of amine in ethyl acetate (*12*). Improvements of this procedure eventually led to a quantitative preparation (*13*) of dimethyl amides of glycine, D, L-alanine, D,L-phenyl alanine and sarcosine from their respective NCA's.

In an excess of Leuchs' anhydride the amine formed from NCA by the action of another amine reacts again with a molecule of NCA and produces a dimer possessing a terminal amine group. Repetition of these reactions leads, therefore, to amine-terminated polypeptides. Hence, the "normal" or "simple" amine-propagated polymerisation of Leuchs' anhydrides is described by the overall reaction,

$$\sim\!\sim CO \cdot CH(R) \cdot NH_2 + (NCA) \rightarrow \sim\!\sim CO \cdot CH(R) \cdot NH \cdot CO \cdot CH(R)NH_2 + CO_2 .$$

Waley and Watson (*14*) were the first to report a kinetic study of this process. They initiated the polymerisation of sarcosine NCA by a pre-formed low molecular weight polysarcosine possessing a terminal amine group. This type of initiation, as those writers pointed out, simplifies the kinetics of the reaction, since the process is reduced to the propagation step only, i. e. one deals with the growth of a living polymer fed by a monomer. The rate of propagation was determined by the increase in CO_2 pressure at constant volume. This method leads, however, to some complications, viz. as the CO_2 pressure rises the degree of carboxylation of the amine to the substituted carbamic acid becomes higher. The consequences of this reaction will be considered later (see p. 9).

The terminal amino groups are not destroyed by the reaction as shown by their titration at the onset of the polymerisation and after its completion. This, therefore, indicates the absence of termination, or a negligible termination, under the conditions maintained in the experiments of WALEY and WATSON.

The absence of termination and of chain transfer is also demonstrated by the molecular weight of the resulting polymer, its number average degree of polymerisation being given then by the simple relation,

$$\overline{DP}_n = p + \text{(monomer supplied)}/\text{(initiator)}_0 ,$$

in which p denotes the degree of polymerisation of the initiating oligomer. Furthermore, in the absence of termination and for an initiation which is not slower than propagation, the resulting product should have a narrow, Poisson-type, molecular weight distribution — a conclusion verified by WALEY and WATSON (14) for the polysarcosine prepared by their technique and subsequently confirmed by FOSTER and OGSTON (15) and by POPE et al. (16). Similar observations were reported for other polypeptides prepared from the respective NCA's by the action of primary amines, e. g. for poly-γ-benzyl-L-glutamate resulting from n-hexyl amine initiation in dimethyl formamide (17). Finally, the lack of termination is proved through the preparation of block-polymers. For example, such a polymer was obtained (13) by initiating polymerisation of D,L-phenyl alanine NCA by a polypeptide formed from sarcosine NCA. Numerous examples of other block-polymer prepared by this technique have been reported since.

A detailed examination of the kinetics of amine-initiated polymerisation of sarcosine led WALEY and WATSON (14) to postulate a reversible addition of amines to NCA, viz.

$$
\underset{CH_3N\text{——}CO}{\overset{CH_2\text{—}CO}{\big\backslash}}\!\!O + HNR_1R_2 \rightleftharpoons
\underset{CH_3N\text{——}CO}{\overset{CH_2\text{—}\underset{\displaystyle |}{\overset{\displaystyle NR_1R_2}{C}}\text{—}OH}{\big\backslash}}\!\!O
$$

The intermediate adduct may be represented, perhaps more realistically, as a resonating zwitter-ion (*18*),

$$
\begin{array}{c}
R_1R_2NH^+ \\
| \\
CH_2\!-\!C\!=\!\!=\!\!O \\
\Big\rbrace\,O^{\ominus}\;, \\
CH_3N\!\!-\!\!-\!\!-\!\!C\!=\!\!=\!\!O
\end{array}
$$

since this form accounts better for the low entropy of activation of the addition process (see p. 11). The decomposition of such an intermediate into a relatively unstable carbamic acid, or its zwitter-ion,

$$
\begin{array}{ccc}
CH_2\cdot CO \cdot NR_1R_2 & & CH_2\cdot CO \cdot NH^+R_1R_2 \\
| & \text{or perhaps} & | \\
CH_3N \cdot COOH, & & CH_3N\!\!-\!\!COO^-,
\end{array}
$$

was demonstrated by BAILEY (*19*) who isolated and identified these compounds. As was shown by him and by others, see e. g. ref. (*31*), these acids rapidly decarboxylate into the amides of the respective amino acids.

The detailed sequence of steps participating in the reaction of primary and secondary amines with NCA's may now be represented by the following scheme (*20*):

$$
\begin{array}{c}
R_1R_2C\!\!-\!\!CO \\
| \quad\quad >\!O + H\cdot Base \underset{k_{-1}}{\overset{k_1}{\rightleftharpoons}} \\
R_3N\!\!-\!\!CO
\end{array}
\qquad
\begin{array}{c}
H\!\!-\!\!Base^+ \\
| \\
R_1R_2C\!\!-\!\!CO \\
| \quad\quad >\!O\Big\rbrace^{\ominus} \\
R_3N\!\!-\!\!CO
\end{array}
\qquad (a)
$$

$$
\begin{array}{c}
Base^+\cdot H \\
| \\
R_1R_2C\!\!-\!\!CO \\
| \quad\quad >\!O\Big\rbrace^{\ominus} \xrightarrow{k_2} \\
R_3N\!\!-\!\!CO
\end{array}
\qquad
\begin{array}{c}
R_1R_2C\!\!-\!\!CO \cdot Base \\
| \\
R_3N\!\!-\!\!COOH
\end{array}
\qquad (b)
$$

$$
\begin{array}{c}
R_1R_2C\!\!-\!\!CO \cdot Base \\
| \\
R_3N\!\!-\!\!COOH
\end{array}
\xrightarrow{k_3}
\begin{array}{c}
R_1R_2C\!\!-\!\!CO \cdot Base \\
| \\
R_3NH + CO_2.
\end{array}
\qquad (c)
$$

Isotope labelling again proved (*21*) that the decarboxylation involves only carbon 2 of the oxazolidine-2,5-dione, and studies of the kinetic isotope effect (*22*) demonstrated that this step is not the rate-determining in the overall process. Therefore, the rate of the overall reaction is determined either by (a) — the formation of the complex, or by (b), viz. by opening of the ring between atoms 1 and 5. This conclusion applies also to the "normal" amine-propagated NCA polymerisation.

Although the most frequent interaction of amines with NCA's involves the 5-CO group, the reaction with the 2-CO group is not entirely excluded (23). This much less frequent event leads to the formation of a ureido-acid,

$$
\begin{array}{ccc}
\text{R} \cdot \text{CH—CO} & & \text{R} \cdot \text{CH} \cdot \text{COOH} \\
| \qquad \diagdown & & | \\
\qquad \quad \text{O} + \text{HN R}_1\text{R}_2 \rightarrow & & \\
| \qquad \diagup & & | \\
\text{NH—CO} & & \text{NH} \cdot \text{CO} \cdot \text{NR}_1\text{R}_2,
\end{array}
$$

which cannot propagate the polymerisation of NCA's and therefore such a reaction results in termination of the polymerisation (23). It should be noticed that this termination takes place in *the course* of polymerisation as a result of a wrong mode of monomer addition; it becomes impossible *after* completion of the polymerisation.

As shown by the data collected in Table 1, the extent of ureido acids formation depends on the nature of the attacking amine (24): for more basic amines it seems to be more pronounced. It requires a higher activation energy than the addition to the 5-CO group and, therefore, this type of termination becomes more significant at elevated temperatures, e. g. in the thermal bulk polymerisation of molten NCA's which yields polypeptides virtually devoid of free α-amino groups.

Table 1. *Yield of Ureido Acids in the reaction of amines with NCA's*

Amine	glycine NCA	sarcosine NCA	phenyl alanine NCA
Diethyl	100	—	80
t-Butyl	90	0	60
i-propyl	45	0	35
Ethyl	55	—	10
Dimethyl	low	low	2.5
Phenyl	—	—	0

From the review by KATCHALSKI and SELA [ref. (2)].

Other termination steps are also possible. For example, HANBY et al. (25) demonstrated in polypeptides formed from γ-benzyl-L-glutamate NCA the presence of a stable, terminal pyrolidone ring. The reaction proceeds as follows:

$$
\begin{array}{ccc}
\text{∾CO} \cdot \text{CH——NH}_2 & & \text{∾CO} \cdot \text{CH——NH} \\
| & & | \qquad\qquad | \\
\text{CH}_2 \quad \text{CO} \cdot \text{O} \cdot \text{CH}_2 \cdot \text{Ph} \rightarrow & & \text{CH}_2 \quad \text{CO} + \text{Ph} \cdot \text{CH}_2 \cdot \text{OH}, \\
\diagdown \quad \diagup & & \diagdown \quad \diagup \\
\text{CH}_2 & & \text{CH}_2
\end{array}
$$

and it may take place *during* or *after* completion of the polymerisation. For example, an amine-initiated polymerisation of γ-benzyl-L-glutamate NCA ($M/I \approx 5$) performed in dioxane at room temperature was completed in about half an hour. The product initially contained terminal amino groups, but most of them disappeared within several days. Similar observations were reported by DOTY et al. (26).

The early work of Wessely suggested the occurrence of some side reactions which led to the formation of diketopiperazines (11) or hydantoins (27). These reactions may put an upper limit to the degree of polymerisation which may be obtained in poly-additions of NCA's but it is more probable that these products arise from an unsuccessful initiation (see p. 40) rather than from a termination. Their participation in the process decreases the yield of CO_2 evolved in the polymerisation.

An interesting termination which may account for the excess of carboxylic groups over that of amine groups in the resulting polypeptide was suggested by Sluyterman and Labruyere (28a). The reaction seems to apply to water-initiated polymerisation of glycine and alanine NCA's performed at elevated temperatures and leads to terminal hydantoin rings, viz.

$$\text{\textasciitilde\textasciitilde NH} \cdot \text{CO} \cdot \text{CH(R)} \cdot \text{NH}_2 + \text{OC—CH(R)—NH} \rightarrow$$
$$\underset{\text{O————CO}}{\mid \qquad\qquad \mid}$$

$$\text{\textasciitilde\textasciitilde NH} \cdot \text{CO} \cdot \text{CH(R)NH} \cdot \text{CO} \cdot \text{CH(R)NH} \cdot \text{COOH} \rightarrow$$

$$\text{\textasciitilde\textasciitilde NH} \cdot \text{CO} \cdot \text{CH(R)} \cdot \text{N} \underset{\text{CO—NH}}{\overset{\text{CO—CH(R)}}{\big<\ \mid\ }} + \text{H}_2\text{O}.$$

This mechanism elaborates that proposed by Sela and Berger (23).

4. Initiation of "normal" NCA's polymerisation by other agents

In his first paper, Leuchs (1) reported that on mixing a small amount of water with glycine NCA, an evolution of CO_2 ensues and a resin was formed. This observation was generalised by him and by others, and for a while the polymerisation of NCA's in moist solvents was considered as an attractive method for preparing polypeptides. In fact, in 1947 Woodward and Schramm (28) reported a polymerisation of L-leucine and D,L-phenyl alanine NCA's in moist benzene which supposedly yielded polypeptides having molecular weights exceeding a million. Unfortunately, this claim turned out to be a mistake. The molecular weight of the resulting co-polymer was determined by osmometry in benzene solution, and in this solvent an extensive agglomeration of the investigated polyaminoacids takes place. The analysis of the end groups by Coleman and Farthing (29) and by other investigators (30) proved that the degree of polymerisation of such a polypeptide does not exceed 100.

Reaction with water produces an amino-acid which eventually initiates (28) the apparently "normal" amino-propagated polymerisation of NCA. The initiation and the hydrolysis are substantially slower than the propagation, e. g. at least a 600-fold water excess is required to hydrolyse NCA without inducing any appreciable polymerisation. Hence, the polymerisation in the presence of water behaves kinetically as an autocatalytic reaction. This process is most useful for the preparation of poly-amino acids possessing free terminal α-amino and α-carboxyl groups. The action of alcohols and ammonia probably takes an analogous course — the produced amine-terminated derivatives act as initiators of NCA polymerisation.

5. Detailed kinetic studies of "normal" amino-propagated polymerisation of NCA's

The pioneering work of WALEY and WATSON (14) was soon extended and elaborated by the studies of BALLARD and BAMFORD (20). They showed that some of the complex features of the kinetics of sarcosine NCA polymerisation, which were reported by the former workers, arose from the catalytic action of carbon dioxide. As the reaction progressed, the pressure of CO_2 increased in WALEY and WATSON's reactor, and hence the contribution of the CO_2 catalysis became time-dependent. To avoid the problem of variable CO_2 pressure, BALLARD and BAMFORD developed a technique in which the pressure of CO_2 was kept constant and the rate of polymerisation was then determined by measuring the increase in CO_2's volume.

Following WALEY and WATSON, the polymerisation was initiated in nitrobenzene or o-nitro-anizole by the pre-formed, low molecular weight "living" polypeptides of $DP \approx 4$. Since a disturbing effect of moisture was noticed, a rigorous drying technique for handling the reagents was recommended. In the absence of moisture, reproducible results were obtained with freshly prepared and rigorously purified anhydrides.

The catalytic action of carbon dioxide, which vitiates the kinetics of sarcosine NCA polymerisation was not observed (20) in similar polymerisations of D,L-leucine and D,L-phenyl alanine NCA's. The kinetics of the latter processes showed simple behaviour. The rates were proportional to the monomer concentration over the whole range of conversion, and a set of experiments performed with different initial concentrations of the starting, living oligomer, see Fig. 1, proved the polymerisation to be a first order reaction with respect to growing ends. The absence of termination was again demonstrated by two observations: (1) the concentration of the terminal amino groups, checked by several analytical techniques, was found to be constant during the whole course of the process, and (2) the $\overline{DP}_n$ of the resulting polymer was given by the equation,

$$\overline{DP}_n = p + \text{(total monomer)}/\text{(total amount of initiating oligomer)},$$

p being the degree of polymerisation of the initiating oligomer.

The kinetic results obtained by Ballard and Bamford are sum-marised in Table 2. The perfect linear relation of the $\log M$ with time, when the reaction was initiated by preformed polymer, provides proof

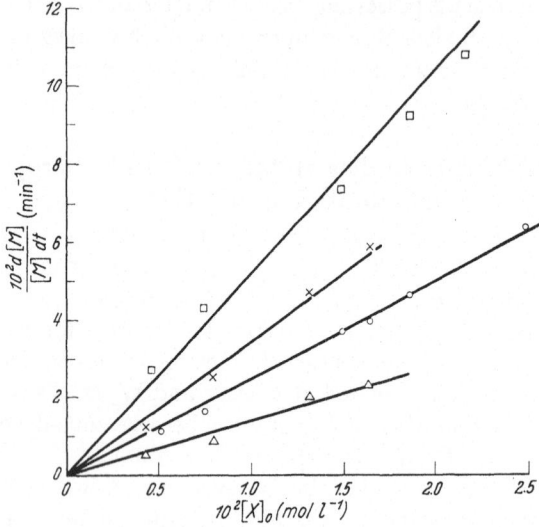

Fig. 1. Polymerization of DL-leucine carbonic anhydride, initiated by preformed polymer. □, solvent, nitro-benzene, 45° C; ×, o-nitroanisole, 45° C; ○, nitrobenzene, 25.2° C; △, o-nitroanisole, 25.0° C. [Reprinted from D. G. H. Ballard and C. H. Bamford: Proc. Roy. Soc. A **223**, 495 (1954) (Fig. 4)]

that all the consecutive steps of the propagation proceed with *the same* rate constant, i. e. $k_{p,j}$ is independent of j for $j \geqq 4$. However, the rate constant of the step,

$$\text{Base} \cdot M + M \rightarrow \text{Base} \cdot M \cdot M \; (j = 1) \,,$$

was found to be three or four times greater than the other $k_{p,j}$'s.

The first order dependence on the monomer and the initiator indicates that the bimolecular formation of the adduct is the rate-determining

Table 2. *The propagation rate constants for the polymerisation of NCA's of* D,L-*leucine and* D,L-*phenyl alanine initiated by the pre-formed polymer*
Rate $= k$ [NCA] [growing ends]

t °C	D, L-leucine $10^2 \, k$ in 1/mole sec		D, L-phenyl alanine $10^2 \, k$ in 1/mole sec
	in nitrobenzene	in o-nitroanisole	in nitrobenzene
25	—	2.38	1.56
25.2	4.17	—	—
45	8.72	5.83	3.50
E, kcal/mole	6.9	8.5	7.7
A, 1/mole sec	$5.0 \cdot 10^3$	$3.8 \cdot 10^4$	$6.0 \cdot 10^3$

From Ballard and Bamford, ref. (20).

step in the polymerisation of D,L-leucine or D,L-phenyl alanine NCA's. Apparently, the opening of the ring and the decarboxylation follow rapidly thereafter. Inspection of Table 2 shows that neither the rate constants of propagation nor their activation energies are drastically affected by the limited change of solvent — a plausible observation in view of their similarities. The very low A factors apparently indicate a greater degree of polarity in the transition state than the initial; this observation therefore justifies the proposed resonating zwitter-ion structure for the adduct (see p. 6).

The simplicity of the kinetics of polymerisation of leucine and phenyl-alanine NCA's contrasts with the complex behaviour of sarcosine NCA. For a constant concentration of growing ends and a constant pressure of CO_2, the propagation was found to be first order in respect to monomer, but the rate increased with the CO_2 pressure as shown in Fig. 2. Moreover, the dependence on the concentration of growing ends was quite complex, and to elucidate this matter it was necessary to investigate the mechanism of the catalytic action of CO_2. This catalysis is clearly demonstrated (20) by Fig. 3 which shows an abrupt decrease in the rate of polymerisation on sudden lowering of the pressure of CO_2 after allowing the polymerisation to proceed for a while at high CO_2 pressure.

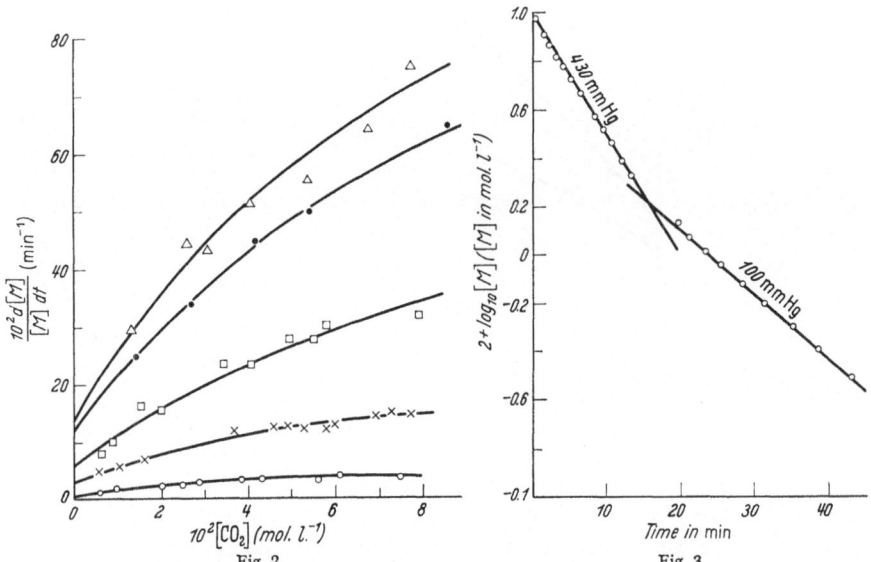

Fig. 2. Polymerization of sarcosine carbonic anhydride in nitrobenzene at 25° C, initiation by preformed polymer. Values of 10^3 $[X]_0$ (mol l⁻¹): △, 1.58; ●, 1.40; □, 1.09; ×, 0.748; ○, 0.400. [Reprinted from D. G. H. BALALRD and C. H. BAMFORD: Proc. Roy. Soc. A 223, 495 (1954) (Fig. 8)]

Fig. 3. Polymerization of sarcosine carbonic anhydride in nitrobenzene at 25° C, initiation by preformed polymer. $[M]_0 = 9.5 \times 10^{-2}$ mol l⁻¹, $[X]_0 = 0.748 \times 10^{-2}$ mol l⁻¹. Pressure was reduced from 430 mm to 100 mm at $t = 15$ min. [Reprinted from D. G. H. BALLARD and C. H. BAMFORD: Proc. Roy. Soc. A 223, 495 (1954) (Fig. 7)]

Studies of Frankel and Katchalski (*31*) proved that carbon dioxide reacts reversibly with amino acids and their derivatives giving the respective carbamic acids, viz.

$$\sim\!\!\text{CO}\cdot\text{CH(R)}\cdot\text{NH}_2 + \text{CO}_2 \ \leftrightharpoons \ \sim\!\!\text{CO}\cdot\text{CH(R)}\cdot\text{NH}\cdot\text{COOH};$$

and, as suggested by Faurholt (*32*), who investigated the interaction of CO_2 with aliphatic amines, further equilibria may be established in the presence of an excess of amine, i. e.

$$\sim\!\!\text{CO}\cdot\text{CH(R)}\cdot\text{NH}\cdot\text{COOH} + \text{H}_2\text{N}\cdot\text{CH(R)}\cdot\text{CO}\!\!\sim \ \leftrightharpoons$$
$$\sim\!\!\text{CO}\cdot\text{CH(R)}\cdot\text{NH}\cdot\text{COO}^-, \text{H}_3\text{N}^+\cdot\text{CH(R)}\cdot\text{CO}\!\!\sim \ \leftrightharpoons$$
$$\sim\!\!\text{CO}\cdot\text{CH(R)}\cdot\text{NH}\cdot\text{COO}^- + \sim\!\!\text{CO}\cdot\text{CH(R)}\cdot\text{NH}_3{}^+.$$

Ballard and Bamford assumed, therefore, that in the presence of CO_2 these compounds were formed in the polymerising mixture. They

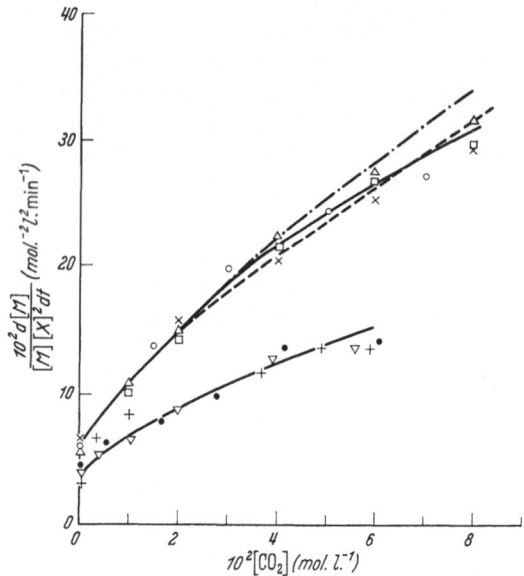

Fig. 4. Polymerization of sarcosine carbonic anhydride in nitrobenzene, initiation by preformed polymer. Values of $10^2\,[X]_0$ (mol l^{-1}): △, 1.58; □, 1.09; ×, 0.748; ○, 0.400; ▽, 1.87; ●, 1.53; +, 1.00. Upper full curve, experimental curve, 25° C; — · — · —, curve calculated from equation (11); — — —, curve calculated from equation (12). Lower full curve, experimental curve, 45° C; curve calculated from equation (11) coinsides with experimental. [Reprinted from D. G. H. Ballard and C. H. Bamford: Proc. Roy. Soc. A 223, 495 (1954) (Fig. 9)]

verified this hypothesis by determining the solubility of CO_2 in the pure solvent and in the solution of amino-terminated polypeptide. The difference was attributed to the "bonded" CO_2, and the results indicated that in 0.2 M solution[1] of the amines the bulk of CO_2 is in the form of the

---

[1] At the concentration of the amines corresponding to the conventional composition of the polymerising mixture the uptake of CO_2 is too low to be reliably determined.

ammonium salts (ion-pairs). The equilibrium constant of the salt formation in nitrobenzene at 25° C was found to be 22 l^2/mole^2 for the oligomer of sarcosine and about 60 l^2/mole^2 for its dimethyl-amide. The latter seems, therefore, to be more basic than the former.

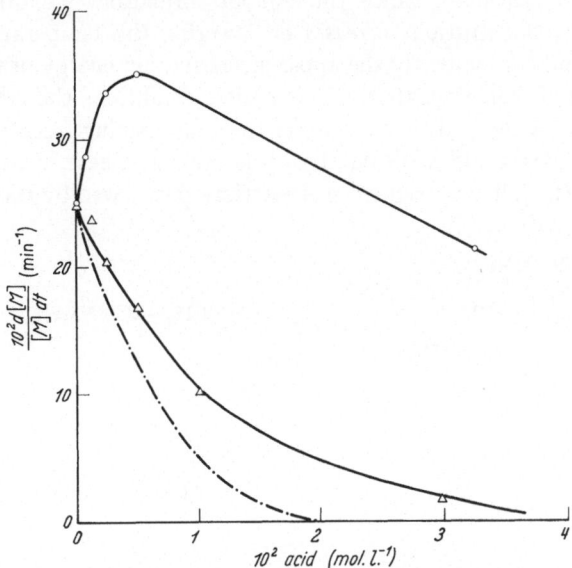

Fig. 5. Polymerization of sarcosine carbonic anhydride in nitrobenzene at 25 °C, initiation by preformed polymer. $[CO_2]$ = 1.05 × 10^{-2} mol l^{-1}; $[X]_0$ = 1.60 × 10^{-2} mol l^{-1}. $\bigcirc$, α-picolinic acid added; $\triangle$, o-nitro-benzoic acid added; — · — · —, calculated curve, based on assumption of nearly complete neutralization of base for acid concentration of 4 × 10^{-2} mol l^{-1}. [Reprinted from D. G. H. BALLARD and C. H BAMFORD: Proc. Roy. Soc. A 223, 495 (1954) (Fig. 10)]

The formation of carbamic acids has a dual effect on the polymerisation: (1) The acids and their salts do not grow — they behave, therefore, as "dormant" polymers which are in dynamic equilibrium with the living ones. Hence, in deriving the correct order of propagation with respect to growing ends, only the concentration of living species should be used for computation. When this is done, calculation shows the propagation of sarcosine NCA polymerisation to be *second* order with respect to *living* ends. This may be seen from the data given in Fig. 4.

(2) The carbamic acids may catalyse the polymerisation, and indeed, in the polymerisation of sarcosine NCA the catalytic effect of weak carboxylic acids, such as α-picolinic, was demonstrated (20). The polymerisation becomes inhibited, however, at high acid concentration, or on addition of a strong acid such as o-nitro-benzoic, since the decrease in the concentration of the base, caused by its conversion into salts, outweighs the catalytic effect of the acid. This, indeed, is seen in Fig. 5.

Apparently, the rate of polymerisation of sarcosine NCA is not entirely determined by the rate of formation of the adduct. The enhancing effect of decreasing temperature on the rate of polymerisation, which was reported by Waley and Watson (14) and confirmed by Ballard and Bamford (20), suggests that the addition of NCA to the growing amine is reversible[1], i. e. k_{-1} (reaction a) is *not* negligible when compared with k_2 (reaction b) (see p. 6). Since the adduct formation is exothermic its stationary concentration *increases* on lowering the temperature of the reaction, and consequently the apparent activation energy of the overall process may become negative[1]. Under these conditions the rate of polymerisation is governed by the rate of the ring opening (reaction b), and apparently bases and acids catalyse this process. These ideas are rationalised by the following sequence of reactions postulated by Ballard and Bamford (20):

$$
\begin{array}{c}
\text{R} \cdot \text{NMe} \\
|\\
\text{CH}_2\!\!-\!\!\overset{}{\text{C}}\!\!-\!\!\text{OH}\!\!-\!\!\downarrow \\
|\quad\quad\diagdown\text{O} + \text{HN(Me)}\text{\tiny\rule[0.4ex]{1.2em}{0.2pt}} \xrightarrow{\ k_2'\ } \\
|\quad\quad\diagup \\
\text{Me} \cdot \text{N}\!\!-\!\!\!-\!\!\overset{}{\text{C}}\!\!=\!\!\text{O}
\end{array}
\qquad
\begin{array}{c}
\text{CH}_2\!\!-\!\!\text{CO} \cdot \text{NMeR} \\
|\\
\text{Me} \cdot \text{N} \cdot \text{COO}^- + \text{H}_2\text{N}^+(\text{Me})\text{\tiny\rule[0.4ex]{1.2em}{0.2pt}}
\end{array}
\qquad \text{(b') *}
$$

and

$$
\begin{array}{c}
\text{R} \cdot \text{NMe} \\
|\\
\text{CH}_2\!\!-\!\!\overset{}{\text{C}}\!\!-\!\!\text{O}\!\!-\!\!\text{H}\!\!-\!\!\downarrow \\
|\quad\quad\diagdown\text{O}\quad\quad| \\
|\quad\quad\diagup \\
\text{Me} \cdot \text{N}\!\!-\!\!\!-\!\!\overset{}{\text{C}}\!\!=\!\!\text{O}\quad\quad\text{O} \\
\uparrow\quad\quad\quad\quad\quad\|\\
\text{\tiny\rule[0.4ex]{3em}{0.2pt}}\!\!-\!\!\text{H}\!\!-\!\!\text{O} \cdot \text{C} \cdot \text{R}_1
\end{array}
\xrightarrow{\ k_2''\ }
\begin{array}{c}
\text{CH}_2 \cdot \text{CO} \cdot \text{NMeR} \\
|\quad\quad\quad\quad\text{H} \\
|\quad\quad\quad\quad| \\
|\quad\quad\quad\quad\text{O} \\
|\quad\quad\quad\quad| \\
\text{Me} \cdot \text{NH} + \text{CO}_2 + \text{OC} \cdot \text{R}
\end{array}
\qquad \text{(b'')}
$$

The former reaction accounts for the catalytic action of a base, and therefore, also for that of a growing polymer, while the latter explains the catalytic action of an acid. Note that an ammonium ion (or ion-pair) may also act as an acid and catalyse the reaction.

The decarboxylation of the zwitter-ions formed by (b') is very rapid (18), and hence, in the absence of acids, the propagation should be second order with base* if the spontaneous decarboxylation of the

---

[1] The negative "activation energy" of polymerisation is sometimes observed in other systems. For example, carbonium-ion polymerisations often proceed faster at lower temperatures than at higher. This is accounted for by a decrease in the rate of termination which has probably a higher activation energy than initiation and propagation. However, the polymerisation of NCA's proceeds without termination, and thus such an explanation cannot apply. The acceleration of some ionic polymerisations on lowering the temperature arises from an increased dissociation of ion-pairs into much more reactive free ions. This explanation again cannot be invoked if the addition (step a) is rate-determining.

* Recent studies of Bamford and Bloch cast doubt on reaction b'. The polymerisation appears to be *not* catalysed by bases, it is nevertheless second order in respect to growing ends. (Private communication from Prof. Bamford.)

primary adduct is very slow and k_2'[base] $\ll k_{-1}$. On the other hand, the kinetic relation, $k_{-1} \ll k_2'$[base], seems to apply in the polymerisation of D,L-leucine or D,L-phenyl alanine, and then the rate of propagation is determined by the rate of formation of the primary adduct. In such a process no catalysis by acids or bases may be observed, and the overall activation energy of the polymerisation must be positive in agreement with experimental findings. We see, therefore, that the change in the sign of the inequality accounts for the different behaviour of sarcosine NCA on the one hand and D,L-leucine or D,L-phenyl alanine NCA's on the other.

The detailed mechanism of sarcosine NCA polymerisation may be radically modified by a change of solvent. For example, in dimethyl formamide this reaction is first order with respect to growing ends [see e. g. ref. (44)] instead of being second order as in nitrobenzene. The higher acidity of the former solvent reduces the basicity of the dissolved amine and, therefore, destroys its catalytic action. This effect influences also the equilibrium between the amine and the dissolved CO_2. In nitrobenzene, the carbamic acid produced is associated with the free amine forming the respective ammonium salt (ion-pairs), whereas in the more acidic dimethyl formamide it exists as an un-ionised acid.

The uniformity of the mechanism for these apparently diverse types of polymerisation is the most appealing feature of the proposed scheme.

The relatively low k_1 values observed in the polymerisations of D,L-leucine and D,L-phenyl alanine NCA's are caused by the low basic strength of the amines formed in these processes. The respective k_1's are probably further decreased by the steric hindrance caused by the bulkiness of the side chains of leucine and phenyl alanine.

The inefficiency of water and of amino-acids in initiating the NCA polymerisation is now easily comprehended. The reaction has to be initiated by a base, whereas the amino-acids exist mainly in their zwitter-ion form. Hence, the initiation must be slow since it involves the non-ionised amino acid present only at low concentration. As the peptide grows in length, the equilibrium between the zwitter-ion and its non-ionised form shifts in favour of the latter, and thus the rate of growth increases. This phenomenon contributes towards the autocatalytic behaviour of this polymerisation.

To recapitulate, the mechanism of the "normal" amine-propagated polymerisation of NCA's assumes the initiation and propagation by primary or secondary amines which act as nucleophiles and add to the 5-CO group of the NCA. The rates of initiation and of propagation increase with the basicity of the respective, initiating or propagating amines, e. g. initiation of D,L-phenyl alanine by the weakly basic p-chloroanaline is 600-times slower than its propagation (30). The presence of bulky substituents in the amine may hamper the rate of their addition and the significance of this will be seen later (p. 38). For the sake of comparison the rates of polymerisation of various NCA's are given in Table 3.

In principle, the addition is reversible and, if the rupture of the $C(5)-O(1)$ bond of NCA adduct is sufficiently slow, an equilibrium may be established between the adduct, the monomer and the initiating or propagating base. The polymerisation of sarcosine NCA seems to be an example.

Table 3. *Rates of amine-initiated polymerisations of NCA's in solution*

NCA of	Solvent	Initiator	$T°C$	$k*) \times 10^2$ l/moles sec	Ref.
D,L alanine	Dioxane	n-hexylamine	35	4.3	33
D,L-α amino n-butyric acid	Dioxane	n-hexylamine	35	1.7	33
glycine	Dioxane	n-hexylamine	35	34.	33
glycine	DMF	diethylamine	25	60.	34
γ-benzyl-1-glutamate	Dioxane	n-hexylamine	35	3.9	33
γ-benzyl-1-glutamate	Dioxane	n-hexylamine	34	4.4	—
γ-benzyl-1-glutamate	Dioxane	n-hexylamine	25	3.2 (0.6)	17
γ-benzyl-1-glutamate	Benzene	diethylamine	25	41.6	34
γ-benzyl-1-glutamate	Nitrobenzene	n-hexylamine	25	11. (2.1)	17
γ-benzyl-1-glutamate	Nitrobenzene	diethylamine	25	7.	34
γ-benzyl-1-glutamate	Chloroform	n-hexylamine	25	30. (4.5)	17
L-leucine	Dioxane	diethylamine	25	1.2	34
L-leucine	Benzene	diethylamine	25	33.	34
D,L-leucine	Dioxane	n-hexylamine	35	0.6	33
D,L-leucine	Nitrobenzene	Preform. polym.	25	4.2	20
D,L-leucine	o-nitro-anizole	Preform. polym.	25	2.4	20
D,L-phenyl alanine	Benzene	diethylamine	25	16.6	34
D,L-phenyl alanine	Dioxane	n-hexylamine	35	1.2	33
D,L-phenyl alanine	Nitrobenzene	Preform. polym.	25	1.6	20
D,L-phenyl alanine	DMF	diethylamine	25	5.8	34
L-proline	DMF	diethylamine	25	75.	34
sarcosine	Nitrobenzene	Preform. polym.	25	25.	14
sarcosine	Acetophenone	Preform. polym.	25	26.	14
D,L-valine	Dioxane	n-hexylamine	35	0.2	33

k is calculated from the equation Rate $= k$ [initiator] $[M]$. From Review by Katchalski and Sela, ref. (*3*).

---

* Some of these polymerisations are heterogeneous, viz. the polymer precipitates as it is formed. Therefore, some of the quoted rate constants may be erroneous.

If the fission of the C(5)—O(1) bond is much faster than the dissociation of the adduct into its original components, the bimolecular formation of the adduct between the monomer and the growing amine then governs the propagation. This is observed in the amine-initiated polymerisation of D,L-leucine or D,L-phenyl alanine NCA in nitrobenzene, or in propagation of γ-benzyl-L-glutamate NCA in dimethyl formamide. Catalysis of the fission cannot affect the rate of those polymerisations for which the adduct formation is the rate-determining step. However if the equilibrium between the NCA, the amine and the adduct is established, catalysis of the fission enhances the polymerisation and then weak acids and bases may act catalytically. The self-catalysis of the growing amine, if it occurs, leads to second order dependence of the propagation on the concentration of growing ends.

Termination is negligible although not entirely eliminated. A "wrong" monomer addition gives, e. g., urea derivatives which cease to grow. The lack of termination leads to the relation $\overline{DP}_n$ of the polymer = Monomer/Initiator, and for a not too slow initiation the resulting product has a Poisson molecular weight distribution.

The proposed mechanism applies to the N-substituted as well as the non-N-substituted NCA's, and in view of the nucleophilic character of the growing ends, the reactions may be classified as an anionic polymerisation.

6. Deviations from the scheme of "normal" polymerisation

Further studies of NCA polymerisation initiated by primary or secondary amines, as well as by other initiators, showed that the mechanism summarised above needs some modification. BLOUT and ASADOURIAN (35) reported that the polymerisation of γ-benzyl-L-glutamate NCA initiated by n-hexyl-amine in dioxane gives two types of polypeptides, recognised today as the helical and the random coil forms, the amount of the latter type of polypeptide decreasing on increasing the M/I ratio. It appears that the molecular weight of the polymer determines the relative proportions of these forms, and the material of high molecular weight seems to contain the helical form only. In the same year some unusual kinetic features of this polymerisation were described (36). As shown in Fig. 6, taken from the LUNDBERG and DOTY paper (17), a slow first order reaction appears to be replaced in the later stages of polymerisation by a fast and also first order reaction, the rather abrupt change in the rate seems to occur when the average DP of the product exceeds the value of 8—10. Moreover, the molecular weight distribution of the resulting polymer was found (26) to be very broad, a fact established for the polymers having molecular weights of about 20,000 or less by applying the Archibald approach-to-equilibrium technique in ultracentrifugation. Had the reaction involved a termination, the anticipated narrow, Poisson-type molecular weight distribution would be broadened; however, as shown by the calculations of Katchalski et al. (37), the termination could *not* account for the substantial broadening observed by Doty (26).

In searching for the solution to these puzzling features of the polymerisation, it was suggested (36) that the polymer starts to grow slowly

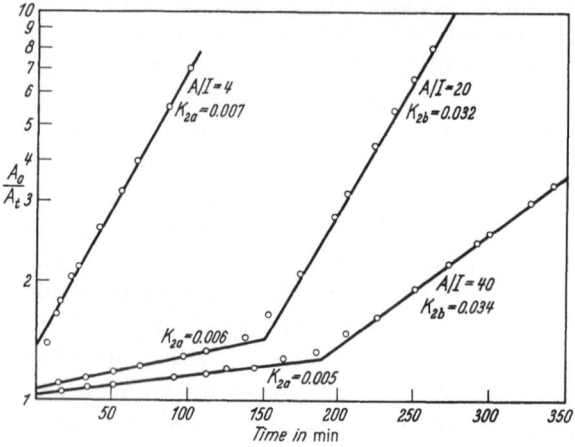

Fig. 6. The polymerization of γ-benzyl-L-glutamate NCA in dioxane at 25° C, initiated by n-hexyl-amine. The values of [NCA] initially and at time t are denoted by A_0 and A_t respectively, "1" is the initial concentration of n-hexylamine. k_{2a}, k_{2b} are the apparent propagation coefficients, in mole^{-1}l sec^{-1}. [Reprinted from paper by R. D. Lundberg and P. Doty: J. Am. Chem. Soc. 79 3961 (1957)]

as a random coil and eventually continues its growth rapidly, after acquiring a helical structure which becomes stable for DP greater than 8–10. The merits of this most interesting suggestion are discussed in section 11 (see p. 54), but at this juncture it should be emphasised that the reported molecular weight distributions, as well as the kinetics of these polymerisations, do not conform with the simple mechanism outlined in the preceding section and demand, therefore, new approaches to the problem.

The problem of two-stage propagation led to an instructive argument. In a brief communication, Ballard and Bamford (38) disputed Doty's findings. They noticed that γ-benzyl-L-glutamate NCA cannot be rigorously purified by sublimation because of its low volatility. The acceleration of polymerisation, they claimed, is observed when the prepared anhydride requires several crystallisations for its purification, while a compound synthesised by an improved procedure, requiring only one rapid crystallisation for its purification, polymerised over the whole range of conversion in a perfectly normal way — the rate being described by a single first order reaction. Moreover, contrary to Doty's observations (17), BALLARD and BAMFORD (38) insisted that the rate of polymerisation of "pure" NCA is the same whether the reaction has been initiated by n-hexylamine or by a preformed polymer with a degree of polymerisation up to 15.

In their reply, DOTY and LUNDBERG (39) question the purity of Bamford's samples, which, they pointed out, were not recrystallised from methylene chloride at $-30°$ C — a procedure considered by BLOUT as essential to ensure the purity of NCA. They correctly stressed that the "simple" kinetic scheme cannot account for a broad molecular weight distribution produced by a two-stage polymerisation. This result seems to be indisputable, e. g. a sample produced by such a polymerisation had $\overline{DP}_n = 20$ and $\overline{DP}_w = 170$.

It has been admitted in later publications by Bamford's group, e. g. in ref. (40), that acceleration *is* observed in the above polymerisations. Samples of NCA's were exchanged between Blout's and Bamford's groups, and the purity of both were found to be satisfactory. Polymerisation of both samples led to acceleration, although the effect observed by Bamford's group was much less dramatic than that reported by DOTY (increase in the rate by a factor of 2 rather than by 5). However, a new problem was raised — Bamford's group questioned the homogeneity of the solution and attributed the acceleration to a gel formation.

The problem of heterogeneity of the reaction will be considered in a later section (see p. 59). The writer wishes to stress, however, that this controversy has been reviewed in so much detail in order to

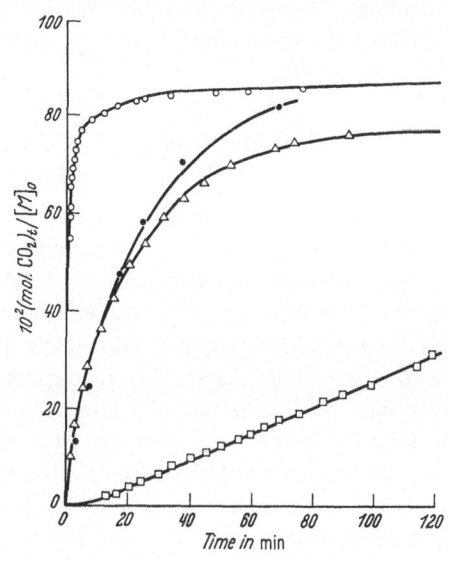

Fig. 7. Conversion-time curves for polymerization of NCA's (0.224 mole 1^{-1}) with amines (0.015 mole 1^{-1}) at 25° C in N,N-dimethylformamide. ○ γ-Ethyl-L-glutamate NCA-di-isopropylamine; △ γ-Ethyl-L-glutamate NCA-n-hexyl-amine; □ Sarcosine NCA-di-isopropylamine; ● Sarcosine NCA-n-hexylamine. [Reprinted from paper by C. H. BAM-FORD, and H. BLOCK: Polyamino Acids, Polypeptides and Proteins, p. 65, Wisconsin University Press 1962 (Fig. 7)]

impress upon the reader the difficulties and pitfalls of research when the results are so extremely susceptible to traces of impurities and to slight changes in experimental conditions.

Another anomaly which calls for revision of the "simple" mechanism of amine-initiated polymerisation of NCA was first reported by BLOUT and KARLSON (49), and stressed recently by BAMFORD and BLOCK (18). As seen from Fig. 7, taken from their paper, the polymerisation of γ-ethyl-L-glutamate NCA initiated by di-iso-propyl amine is *faster* and yields higher polymer than the reaction performed under identical conditions but initiated by n-hexylamine. The "normal" initiation by both amines is not *slower* than the propagation and, therefore, the rate constant of polymerisation yielding peptides of DP 10 or higher should be *independent* of the initiator's nature if the "simple" mechanism is valid (see e. g. the data listed in Table 3).

Evidence provided by BLOUT and KARLSON (49) and by BAMFORD and BLOCK (18) implies, therefore, that there are at least two modes of *propagation* even in a polymerisation initiated by primary or secondary amines. The evidence for other modes of propagation is apparent from studies of polymerisations initiated by tertiary amines and other aprotic initiators. We shall therefore review these reactions first, and then continue the discussion of possible mechanisms of these most intriguing polymerisations.

7. Polymerisation of NCA's initiated by tertiary amines and other aprotic bases

Tertiary amines were amongst the first initiators of NCA polymerisation which had been described in the literature and it seems that the polymerisation of all the known NCA's may be accomplished by their action. WESSELY (11) reported in 1925 that glycine and phenyl alanine NCA's are readily polymerised in pyridine at ambient temperatures, and in the following paper (12) he reported a similar polymerisation of sarcosine NCA. The polypeptides produced by this initiator apparently formed *cyclic* polymers since no terminal end groups could be detected (41). It is significant that appreciable quantities (a few %) of 3-acetichydantoin derivatives were found in the polymers formed from glycine and phenyl alanine NCA's but *none* was detected in the polymerised sarcosine NCA (12). This evidence suggests that the mechanisms of polymerisation initiated by aprotic bases may be different for the non-Nsubstituted NCA and the N-substituded anhydrides.

In the early years of these studies, some controversy arose as to whether traces of water are necessary for the initiation induced by tertiary amines. The initiation by primary and secondary amines has been explained by postulating transfer of a labile proton to the monomer. Such protons are not available in a tertiary base, and therefore COLEMAN (42) suggested a co-catalysis by traces of water which would supply the protons. However, the elaborated and painstaking studies of Wessely's group (41) demonstrated that water is unnecessary for the pyridine initiation, and more recent studies of Bamford's group (*43, 44*) have confirmed this fact.

Another controversy has been created by the problem of whether tertiary amines genuinely initiate the polymerisation of the N-substituted NCA's. WESSELY (*11, 12*) found pyridine unable to initiate the polymerisation of N-phenyl glycine NCA, but he reported the polymerisation of sarcosine in rigorously-dried pyridine (*41*). This observation was questioned by Bamford's group (*43, 44*) — they claim explicitly that carefully purified sarcosine NCA is *not* polymerised by tertiary amines. It is even more difficult to decide whether L-proline NCA, another N-substituted anhydride, or any of its derivatives may be polymerised by aprotic bases. The extremely high reactivity of these monomers makes the experimental results disputable. Pyridine-initiated polymerisations of these NCA's were observed by Katchalski's group (*45*), whereas Bamford's team (*18, 46*) reports only an extremely slow polymerisation of L-proline NCA in the presence of a tertiary base, although a very rapid polymerisation ensues on addition of 3-methyl hydantoin to a slowly polymerising mixture. It is difficult to visualise an adventitious inhibitor which could slow down the polymerisation studied by BAMFORD, whereas it is plausible to assume the presence of some adventitious impurities which enhanced the reaction studied by other workers. Therefore, the results of Bamford's group appear to be genuine, and the N-substituted NCA's might not be readily polymerised by tertiary amines, if all proton-donating impurities are rigorously excluded. However, Bamford's conclusion should not be generalised, and it might not apply to aprotic bases stronger than tertiary amines (see, e. g., p. 47).

The NCA's polymerisation initiated by tertiary amines and by other aprotic bases such as sodium hydroxide, sodium methoxide, triphenyl methyl sodium, etc., is of the greatest significance since this method led to truly high-molecular weight synthetic polypeptides. Such materials are most valuable as models for extensive studies of physical and chemical properties of proteins. Their synthesis was reported in 1954 by BLOUT, KARLSON, DOTY and HARGITAY (*47*) who polymerised γ-benzyl-L-glutamate NCA in dry dioxane, or its mixture with tetrahydrofuran, using methanol solution of sodium hydroxide as the initiator. The polymerisation was carried out at 25° C and the reaction was completed in about 4 hours. The molecular weight of the product was found to exceed 100,000 and, in fact samples having M. W. 350,000 were reported (*48*). The determination of molecular weights was accomplished by the light-scattering technique, employing the Zimm plots over the angular range 30—135°. The absence of polymer association was definitely established since the results obtained in dichloroacetic acid agreed with those found in chloroform-formamide solution. Moreover, the reduced viscosity varied linearly with concentration down to the lowest measurable dilutions.

Rigorous purification of NCA and meticulous exclusion of any traces of moisture, hydrochloric acid, amino-acids etc., appear to be imperative for a reproducible preparation of really high molecular-weight materials (47), and further improvements (49) eventually led to polypeptides having molecular weight over 1,000,000.

8. Conditions leading to high molecular-weight synthetic polypeptides

The discovery of high molecular-weight synthetic polypeptides came as a result of systematic studies by BLOUT and his co-workers of the conditions which limit the degree of polymerisation of products formed by polymerisation of NCA. Contrary to previous beliefs, it was found that the polymerisation initiated by n-hexyl amine in dry dioxane yields polypeptides of higher molecular weights than expected from the monomer-to-initiator ratio, provided $M/I < 100$. It should be stressed again that this observation demonstrates that some additional modes of propagation contribute to the reaction, i. e. the "simple" poly-addition mechanism cannot account *in toto* even for the polymerisation initiated by primary amines in dioxane.

The increase in the M/I ratio above 50 leads to only a small increase in the molecular weight of the polymer which remains virtually constant when M/I exceeds 400 or 500. Hence, some inherent termination steps limit further growth of the polymeric molecule and, therefore, the n-hexyl amine initiated polymerisation cannot give products of $\overline{MW}$ greater than 100,000, or $\overline{DP}$ higher than 450. Searching for a method which would give still higher molecular-weight polypeptides, BLOUT explored the action of other initiators and methanol solution of sodium hydroxide was the first which produced the desirable result. As shown in Fig. 8, taken from the paper by BLOUT and KARLSON (49), for an M/I ratio of about 50 the molecular weight of the product was twice as high for the polymer initiated by NaOH than for that formed through initiation by n-hexyl amine. A limiting value of about 500,000 was reached when $[M]/[\text{NaOH}]$ exceeded 500.

These results stimulated investigation of other alkali initiators. For the ratio $M/I \approx 50$, sodium methoxide yielded a polymer having $\overline{MW}$ in excess of 100,000, and products having $\overline{MW}$ over 600,000 were obtained when M/I was greater than 1000. Similar results were reported for sodium borohydride initiation. Most unexpectedly, polymerisation initiated by diethyl amine in dioxane solution produced a polypeptide having a constant molecular weight of about 200,000 independent of the value of the M/I ratio which was varied from 50 up to 1000. Initiation by triethyl amine produced an even higher degree of polymerisation

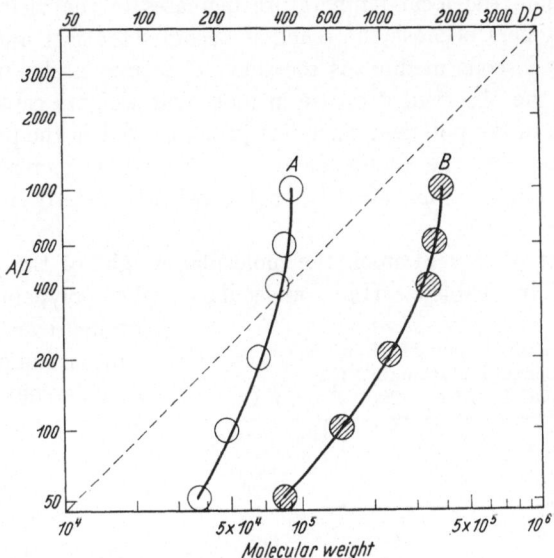

Fig. 8. Poly-γ-benzyl-L-glutamates from polymerizations in dioxane solution; with n-hexylamine initiator, A; with sodium hydroxide initiator, B; degree of polymerization (D. P.) as a function of anhydride-initiator ratio (A/I). The molecular weights were obtained from reduced specific viscosities at concentrations of 0.2% in dichloroacetic acid. [Reprinted from paper by E. R. BLOUT and R. H KARLSON: J. Am. Chem. Soc. **78** 94 (1956) (Fig. 1)]

approaching 4500 ($\overline{MW} \approx 1,000,000$). Since the tertiary amine initiation was known for a long time, it is most surprising that synthesis of extremely high molecular weight polypeptides was not reported earlier. The correlation of the initiator's nature with the molecular weight of the polypeptide is summarised in Table 4.

All the polymerisations discussed above were found to proceed much faster (47, 49) than those previously observed. For concentrations of monomer ranging from 0.5 to 5% the molecular weights of the produced polypeptides appear to be constant; however, they *decline* on further increases of monomer concentration. Raising the temperature of polymerisation above 30° C also had an unfavourable effect upon the

Table 4. *Effect of the initiator on M. W. of the polypeptide*

Monomer: γ-benzyl-L-glutamate NCA; $M/I = 5$
Solvent: dioxane; $T = 25°$ C

Initiator	Molecular Weight $\times 10^{-3}$, from $[\eta]$ in dichloroacetic acid
n-Hexyl amine	15
Diethyl amine	83
Triethyl amine	280
NaOH	31
NaOCH$_3$	29.5
NaBH$_4$	32.5

From BLOUT and KARLSON; ref. (49).

molecular weight of the polymer, and this might explain the strange effect of the monomer's concentration. It may be that in a more concentrated solution, the local temperature increases in the vicinity of the growing polymers because the reaction becomes too fast and the heat transfer in a viscous medium is too slow. The same explanation might account for the observed decrease in molecular weights caused by the precipitation of the polymer, since it is probable that in the precipitated phase, swollen by the monomer, its concentration increases in the vicinity of growing ends (for further discussion of heterogeneous polymerisation see p. 59).

The effect of solvent upon the molecular weight of the product is summarised in Table 5. High molecular-weight polypeptides *were* obtained even in benzene solution, indicating that the polar nature of the solvent is not imperative for these processes. This is puzzling, since there are good reasons to believe that some ionic species participate in these polymerisations and obviously a hydrocarbon solvent does not favour their formation (see, however, p. 48). On the other hand, it is possible that the growing end of a high molecular weight polymer is surrounded by its polar chain. Thus, its immediate environement may be polar even when the bulk of the solvent is non-polar.

Table 5. *Solvent effect upon M. W. of the polypeptide*
Monomer: γ-benzyl-L-glutamate NCA
Initiator: $NaOCH_3$; $M/I = 200$; $T = 25°\,C$

Solvent	$[M]$ grms/100 cc	Molecular Weight $\times\,10^{-3}$ from $[\eta]$ in dichloroacetic acid
Dioxane	2	365
Anisole	2.5	345
Benzene	1	340
Chloroform	2.5	290
Chlorobenzene	2.5	230
Dioxane + 0.1% of hexaldehyde	5	170
		135
Ethyl acetate	2.5	134
Nitrobenzene	2.5	124
Acetonitrile	2.5	115
Dimethyl formamide	5	83
Nitromethane	2.5	59
Methanol	2.5	12
Dioxane + 1% of hexaldehyde	5	none

From Blout and Karlson; ref. (*49*).

Data quoted in Table 5 illustrate also the effect of some impurities on the molecular weight of the product. The presence of 0.1% of hexaldehyde in the dioxane solution was sufficient to reduce the molecular weight of the product by a factor of 2, and only a low molecular-weight oligomer, not precipitated by ethanol, was produced when the proportion of the aldehyde increased to 1%. The presence of acetone had no influence on the product. It is possible that the impurities present in the reacting mixture, or formed by some side reactions, impose an upper limit on the maximum molecular weight of the polypeptide formed in this system.

BLOUT and KARLSON (49) tried to explain the high DP values of the polymers produced at relatively low ratios M/I by assuming that the interaction of the initiator with the monomer involves two or more simultaneous reactions, of which only one gives a product leading to the high polymers while the products of the other reactions do not grow, i. e.

$$\text{Initiator} + \text{Monomer} \nearrow \overset{\text{Product I} \;\rightarrow\; \text{Polymer}}{\searrow} \;\text{Product II} \rightarrow \text{no Polymer}.$$

Such a scheme might account for the inequality $DP > M/I$. However, it demands also, contrary to the observations, a proportionality between DP and M/I.

Further peculiarities of the strong base initiated polymerisation were revealed by the work of IDELSON and BLOUT (50). Polymerisation was found to ensue each time when the monomer, γ - benzyl - L - glutamate NCA, was added to a solution containing products resulting from a previously - completed polymerisation initiated by sodium methoxide in dioxane. Each post-polymerisation resulted in a quantitative conversion of the added monomer into polymer and, for a constant amount of monomer added at each stage of the process, a constant molecular weight was found for the

Table 6. *Post-polymerisation of γ-benzyl-L-glutamate NCA. The reaction was initiated by sodium methoxide in dioxane solution at constant monomer concentration (4%)*

Exp.	Total M/I	Added M/I	$\overline{DP}_w$
1	20	—	650
	40	20	750
2	20	—	550
	40	20	650
	60	20	650
	80	20	650
	100	20	700
3	100	—	1100
	200	100	1100
4	400	—	2150
	800	400	2250
5	1000	—	3350
	2000	1000	3500

From IDELSON and BLOUT; Ref. (50).

product, proving that this phenomenon cannot be explained by the growth of living polymers. Typical results of such experiments are shown in Table 6.

IDELSON and BLOUT also demonstrated that the molecular weight of the product formed in a single-stage polymerisation did *not* increase as the conversion of the monomer rose from 25% to 100%. This is shown in Table 7 taken from their paper. Since virtually all the monomer was converted into polymer, we see again that the concept of living polymers cannot account for this process.

This result, and the occurrence of post-polymerisation, are highly significant. They indicate:

(1) A continuous creation and destruction of growing chains probably leading to their stationary concentration in the course of polymerisation.

(2) A similar kinetic order in respect to monomer for propagation and termination, since the molecular weight of the formed polymer remains nearly constant in spite of varying concentration of the monomer. The termination arises perhaps from a chain transfer to monomer or a wrong monomer addition which prevents further propagation (23).

Table 7. *Degree of polymerisation of poly-benzyl-L-glutamate at various conversions*

M/I	NCA consumed %	$\overline{DP}_w$
100	25	1100
	50	1250
	75	1250
	100	1250
200	31	800
	40	1000
	53	1050
	62	1050
	89	1100
	100	1200
400	40	1650
	51	1750
	73	2100
	100	2400
1000	21	2550
	51	3050
	76	3450
	100	3350

From Idelson and Blout, ref. (50).

(3) The initiator, or any product formed from it which in turn initiates the reaction, is virtually not consumed by the polymerisation since the process could be repeated over and over again on each addition of fresh monomer. The following alternative explanations may be considered: (a) the initiating species are regenerated in the course of polymerisation; (b) the initiator's role is to activate the monomer rather than to initiate a growing polymeric chain. On its incorporation into the chain, the activated monomer regenerates the "initiator" or activates another molecule of monomer. Such a mechanism was discussed — it accounts, e. g., for the polymerisation of pyrrolidone (60), and this fruitful idea, adopted for NCA polymerisation, has been advocated and substantiated by Bamford's school (18, 51, 52).

On first inspection, the effect of the M/I ratio on the $\overline{DP}$ of the produced polymer appears to be puzzling. For example, the data shown in Table 6 indicate a higher $\overline{DP}$ if the product results from the addition of *all* the monomer at the onset of the reaction, while its stepwise addition in small portions, albeit in the same total quantity, gives a slightly lower $\overline{DP}$. However, as shown in Table 6, the *initial* concentration of the monomer was kept constant (4%) in Blout's experiments (50), and the increase in the M/I ratio was achieved by *decreasing* the concentration of the initiator. Hence, the results given in Table 6 indicate a slight decrease of $\overline{DP}$ on increasing the initiator's concentration, implying,

e. g., that the initiator or the non-activated monomer affect the termination or chain transfer. Whatever the correct interpretation of these effects, it should be stressed that they are small, and in some systems, e. g. for a polymerisation initiated by diethyl amine or triethyl amine (*49*), the $\overline{DP}$ of the polymer is *independent* of the initiator's concentration. It seems that two or more propagation steps simultaneously contribute to the growth of the polymeric molecule and probably the termination of only one of them, perhaps of the least significant, is affected by the initiator's concentration.

9. Kinetics of NCA polymerisation initiated by aprotic bases

Most of the kinetic studies of NCA polymerisation utilise the evolution of carbon dioxide, the amount of which measures the degree of conversion. In addition to gasometric techniques which operate at constant volume or at constant pressure, a continuous titrating technique was developed by PATCHORNIK and SHALITIN (*52a*). Nitrogen saturated with solvents vapour is bubled through the reacting solution and then passes through the absorbing train. The CO_2 swept by the gas is titrated as absorbed. An alternative method determining the amount of reacted NCA, was developed by IDELSON and BLOUT (*53*). The IR absorption spectra of NCA show two characteristic bands at 1860 and 1790 cm⁻¹ which are attributed to the C=O stretching vibrations. On conversion of the monomer into polymer these disappear, and therefore, the progress of the reaction may be followed by a decrease in the optical density at these wavelengths. An additional band at 1735 cm⁻¹, present in the IR spectrum of γ-benzyl-L-glutamate NCA, arises from the stretching vibration of the carbonyl group of the benzyl ester moiety, and since its intensity is not affected by the polymerisation, it conveniently serves as an "internal standard". The IR spectrophotometric technique, developed along these lines, was applied by IDELSON and BLOUT in their kinetic studies of γ-benzyl-L-glutamate NCA polymerisation, the results being checked by the ordinary gasometric technique. The agreement was most satisfactory.

The basic kinetic features of the "simple" polymerisation of NCA were described and discussed in sections 3 and 4. It has been shown that the "simple" or "normal" polymerisation of the non-N-substituted, as well as of the N-substituted, NCA's proceeds by the same amine propagated mechanism. Here we shall be concerned with the propagation initiated by aprotic bases and similar agents. We shall explore also the problem whether the two types of NCA monomers, the N-substituted and the non-N-substituted, behave similarly in such polymerisations.

Studies of IDELSON and BLOUT revealed an important kinetic feature of the polymerisation initiated by strong bases, namely an initial accelerating period is followed by a pseudo first order reaction (see Fig. 9). It seems that this behaviour accounts for the previously mentioned findings of Lundberg and Doty (17) who reported two first order reactions — a slow one observed at the initial period of the process and a fast one at later stages. The accelerated portion of the conversion curve could easily be mistaken for a slow first order reaction. The initial acceleration, confirmed by other workers, e. g. see ref. (59), results most probably from a relatively slow approach to the stationary state of the

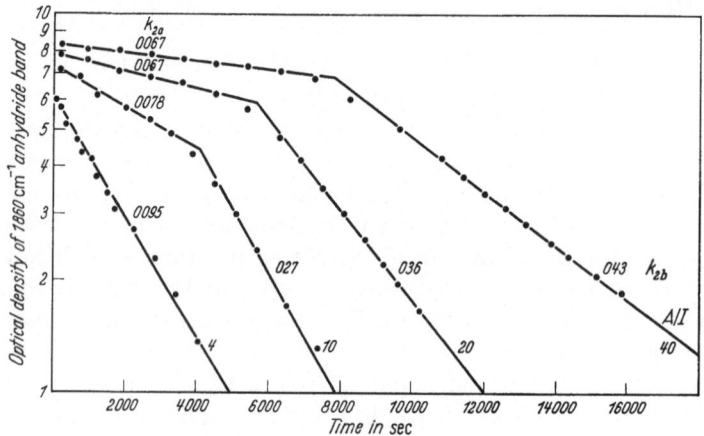

Fig. 9. Plots of optical density of 1860 cm⁻¹ NCA C=O band measured as a function of time during n-hexyl-amine initiated polymerizations of γ-benzyl-L-glutamate-N-carboxyanhydride at A/I 4, 10, 20 and 40. The initial rate constants, k_{2a}, are shown below the symbol. The final rates, k_{2a}, are shown in a horizontal line with the symbol. [Reprinted from paper by M. Idelson and E. R. Blout: J. Am. Chem. Soc. 79, 3948 (1957) (Fig. 2)]

growing species. Such a situation is characteristic for those polymerisation processes in which the rate of initiation and termination are slow compared with the rate of growth.

The high rate of NCA polymerisation initiated by aprotic bases is its most remarkable kinetic feature. This is due to the high rate of propagation since the concentration of growing species appears to be low. It is apparent, therefore, that the mechanism of *propagation* of this reaction differs from that postulated for the primary amine-initiated polymerisation.

Detailed kinetic studies of some aprotic systems were reported by Bamford's group. The kinetics of the polymerisation initiated by sodium salt of dihydrocinnamic acid in N-methyl-formamide exhibits a few interesting features. Typical conversion curves for D,L-phenyl alanine and γ-benzyl-L-glutamate NCA's are shown in Fig. 10 taken from the paper by Ballard and Bamford (54). For a low concentration of the salt, the reaction is self-inhibited, i. e. a relatively rapid polymerisation ceases after a while (curve C) when a substantial fraction of the still unreacted NCA remains in the solution. With a larger salt concentration a complete conversion is observed (curve A). It is apparent that an inhibitor is formed in the process, and BALLARD and BAMFORD (54) proved that the formation of hydantoin-3-acetic acid is responsible for this behaviour. The yield of this acid is larger in the polymerisation of D,L-phenyl alanine than in the reaction of γ-benzyl-L-glutamate.

The initial rate of polymerisation appears to be first order with monomer for a low salt concentration; the order of the reaction increases,

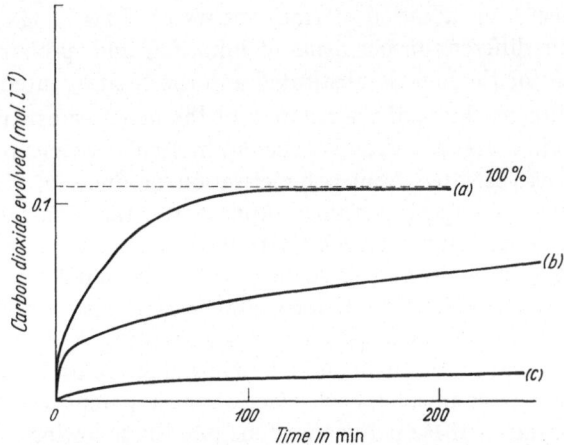

Fig. 10. Conversion-time curves for polymerisation of NCA's in 80% nitrobenzene —20% N-methyl formamide mixture; 25° C. [NCA] = 0.109 mol. l^{-1} initially in all cases. (a) γ-benzyl-L-glutamate NCA [sodium di-hydrocinnamate] = 4.6 × 10^{-3} mol l^{-1}; (b) DL-phenylalanine NCA [sodium dihydrocinnamate] = 4.6 × 10^{-3} mol l^{-1}; (c) DL-phenylalanine NCA [sodium hydantoin 3-acetate] = 9.2 × 10^{-3} mol l^{-1}. 100% evolution of CO_2. [Reprinted from paper by D. G. H. BALLARD and C. H. BAMFORD: Special Publication of Chem. Soc. No. 2, p. 25 (1955) (Fig. 1)]

however, for higher concentrations of the initiating salt. Sarcosine NCA, which is generally *more* reactive than the other two NCA's, reacts very slowly, and probably even this slow reaction is induced by some impurity present in the solvent. This N-substituted NCA was also found to be inert in respect to polymerisation initiated by LiCl in dimethyl forma-mide (44), although this system initiates a rapid polymerisation of the non-N-substituted NCA's.

The kinetics of the relatively fast polymerisation of the non-N-substituted NCA's initiated by LiCl in dimethyl formamide was thorough-ly studied by BALLARD, BAMFORD and WEYMOUTH (44). Addition of this salt enormously speeded, by a factor of at least 1000, the extremely slow spontaneous polymerisation of glycine NCA. (In the absence of salt no reaction was detectable even after a few hours.) Other NCA's behaved similarly, although in the absence of the salt their spontaneous poly-merisations were slightly faster than that of glycine NCA. The initial rate of reaction was found to be proportional to the square of monomer concentration, indicating the combined effect of the monomer on the rate of initiation and propagation.

Traces of water have no effect upon the rate of the reaction. This was demonstrated by changing the method of drying of LiCl and by deliberate addition of small amounts of water to the system containing dry LiCl. In this respect a very different behaviour is shown by the N-substituted NCA's, e. g. the sarcosine anhydride. Although its solution in dimethyl formamide does not react in the presence of dry LiCl, a fast reaction ensues on addition of traces of water. This is again a strong evidence for different mechanisms of initiation and, probably also, of propagation for the non-N-substituted and the N-substituted NCA's

The major products of the reaction of the non-N-substituted NCA's are hydantoin-3-acetic acid and cyclic low molecular weight polypeptide. The latter was isolated from polymerisation products of glycine NCA and identified as a cyclic hexamer. Apparently, the high proportion of salt speeds up the intra-molecular termination (see also p. 48). On the other hand, polymerisation of sarcosine NCA initiated by LiCl *in the presence* of 3-methyl-hydantoin, a co-catalyst is necessary in the absence of water, produced linear polypeptides possessing a terminal amine group on one end of the chain and a hydantoin group probably attached to the other. The significance of this fact will be apparent when we discuss the mechanisms of these polymerisations (see the following section).

The LiCl initiated polymerisation is also inhibited by acid, and hence the formation of hydantoin-3-acetic acid leads to auto-inhibition.

Ballard and Bamford (43) also investigated the kinetics of the polymerisation of D,L-phenyl alanine NCA initiated by tributyl amine in nitrobenzene. This reaction does not ensue[1] in pure tertiary amine (11), but it proceeds rapidly in polar solvents. (For a tentative explanation of this observation see p. 49.) However, the polarity does not appear to be a necessary condition for a rapid reaction since a fast polymerisation initiated by triethyl amine in benzene was observed by Blout (49). It seems, therefore, that the role of solvent in these reactions needs further and extensive examination (see also p. 48). In the tributylamine-initiated polymerisation, sarcosine NCA again proved to be inert (44) under conditions which led to a rapid polymerisation of the non-N-substituted NCA's.

The time conversion curves of the polymerisation initiated by tributyl amine are shown in Fig. 11. They resemble the curves presented in Fig. 10, indicating the basic similarity of the process initiated by tertiary amines, by salts of carboxylic acids and by LiCl in dimethyl

---

[1] It might be desirable to reinvestigate this claim.

formamide. The auto-inhibition was observed at a low base concentration and its cause was traced again to the formation of substituted 3-hydantoinyl-acetic acids. The order of the initial rates with respect to base is slightly lower than one, and slightly higher than one with respect to monomer. At constant monomer concentration (about 1 m/l) the intrinsic viscosity of the final polypeptide increases with increasing M/I ratio, attains a maximum at $M/I \approx 50$ and then decreases. These changes are shown in Fig. 12.

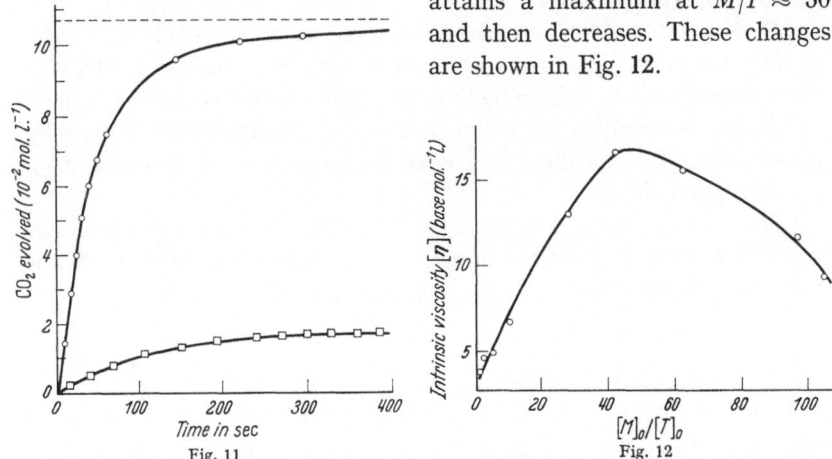

Fig. 11

Fig. 12

Fig. 11. Conversion-time curves for reaction of N-carboxy-DL-phenylalanine anhydride with tri-n-butylamine in nitrobenzene at 25, $[M]_0 = 0.107$ mol^{-1}. O $[T]_0$ (initial total base concn.) $= 0.103$ mol l^{-1}; □ $[T]_0$ $= 1.6 \times 10^{-3}$ mol l^{-1}. Broken line corresponds to 1 mol of CO$_2$. [Reprinted from paper by D. G. H. BALLARD and C. H. BAMFORD: J. Chem. Soc. **1956**, 381 (Fig. 1)]

Fig. 12. Intrinsic viscosities of polymers prepared in nitrobenzene at 25° as a function of $[M]_0/[T]_0$. $[M]_0$ $= 1.04$ mol l^{-1} in all cases. [Reprinted from paper by D. G. H. BALLARD and C. H. BAMFORD: J. Chem. Soc. **1956**, 381 (Fig. 4)]

10. The mechanism of polymerisation initiated by aprotic bases

Any proposed mechanism of polymerisations initiated by aprotic bases must account for the following observations:

(1) The initiation proceeds without proton transfer *to* monomer, and the initiator is not consumed in the reaction.

(2) Polymerisation initiated by aprotic bases proceeds faster and gives products of higher molecular weight than the reaction initiated by primary amines. Hence, the propagation mechanism of the former must differ from that of the latter, its rate constant being higher than that of the "simple" amine-propagated reaction.

(3) Increase in the initiator's concentration enhances the polymerisation, although the concentration of growing species does not necessarily increase.

(4) The degree of polymerisation of the resulting polypeptide may be substantially higher than the M/I ratio. Moreover, for a constant M/I the $\overline{DP}$ depends on the initiator's nature.

(5) The molecular weight distribution of the polymer may be very broad[1], e. g. the ratio $\overline{DP}_w/\overline{DP}_n$ could be as large as 10.

(6) No long-lived polymers are formed, although the polymerisation re-ensues each time on addition of fresh monomer to the polymerised mixture.

(7) Cyclic polypeptides, or polypeptides devoid of terminal amino groups are amongst the products of, at least, some polymerisations initiated by aprotic bases.

(8) Hydantoin-3-acetic acid or its derivatives are frequent byproducts of these reactions. They were not found, however, amongst products of reaction initiated by sodium methoxide (BLOUT, private communication).

(9) It seems that the polymerisation of N-substituted NCA's by aprotic bases takes a different course from the reaction involving non-N-substituted NCA's.

The various mechanisms reported in the literature for these processes will now be reviewed. Some schemes (50) assume initiation in which the initiator's fragments are incorporated into the polymer, e. g.,

$$
\begin{array}{ccc}
\text{R} \cdot \text{CH}\!-\!\text{CO} & & \text{R} \cdot \text{CH}\!-\!\text{CO} \cdot \text{OCH}_3 \\
| \quad \quad \rangle\text{O} + \text{Na}^+\text{OCH}_3^- \rightarrow & | & \xrightarrow[\text{monomer}]{\text{more}} \\
\text{NH}\!-\!\text{CO} & & \text{NH} \cdot \text{COO}^-, \text{Na}^+
\end{array}
$$

$$\text{CH}_3\text{O} \cdot \text{CO} \cdot \text{CH(R)} \cdot \text{NH} \{\text{CO} \cdot \text{CH(R)} \cdot \text{NH}\}_n \cdot \text{CO} \cdot \text{CH(R)} \cdot \text{NH}^-, \text{Na}^+ + (n+1)\,\text{CO}_2 \; [2].$$

Such mechanisms might be questioned since in a recent work GOOD-MAN and ARNON (55) demonstrated the absence of initiator's fragments in the resulting polymer. In the course of their studies, polymerisation of γ-benzyl-L-glutamate was initiated in dioxane by 9-fluorenyl potassium or by radio-active sodium methoxide. The polymer produced by the former initiator was precipitated and examined spectrophotometrically. For M/I ratio of 60 its degree of polymerisation was found from its intrinsic viscosity to be about 180 and its solution did not absorb at $\lambda = 300$ mμ where a strong absorption band ($\varepsilon = 10^4$) of the fluorenyl moiety should appear. On the other hand, all the fluorenyl residues were found in the solution left after precipitation of the polymer.

Similar results were obtained with radio-active sodium methoxide. The activity of the precipitated polymer was only 0.7% of that expected had each polymer molecule contained one methoxy group. Again, virtually all the activity was found in the residual solution.

The absence of initiator fragments in the polypeptide implies regeneration of the initiator in the course of polymerisation or an initiation

---

[1] However, the high molecular weight poly-peptides resulting from initiation by sodium methoxide did not show unusually broad distribution (unpublished results of BLOUT und DOTY).

[2] This equation does not necessarily imply decarboxylation of the carbamate. The salt is stable and the carbamate ion must be neutralised prior to its decarboxylation.

process arising from a transfer of some monomer's fragment, e. g. a proton, to the initiating base.

Two of the published mechanisms visualise regeneration of the initiator. WIELAND (56, 57) suggested that tertiary amines convert NCA molecules into zwitter-ions, i. e.

$$\begin{array}{ll} R \cdot CH\!-\!CO & R \cdot CH\!-\!CO \cdot {}^{+}NR_3' \\ \quad\Big| \qquad \diagdown & \quad\Big| \\ \quad\Big| \qquad \diagup O + NR_3' \rightleftharpoons & \quad\Big| \\ NH\!-\!CO & NH\!-\!COO^{-}. \end{array}$$

Their interaction then leads to a linear anhydride which on decarboxylation gives a peptide bond, viz.

$$\begin{array}{ll} R \cdot CH \cdot CO \cdot {}^{+}NR_3' & {}^{-}O \cdot CO \cdot NH \\ \quad\Big| \qquad\qquad + & \quad\Big| \qquad\qquad \rightarrow \\ {}^{-}O \cdot CO \cdot NH & R \cdot CH \cdot CO \cdot {}^{+}NR_3' \end{array}$$

$$\begin{array}{l} R \cdot CH \cdot CO \cdot O \cdot CO \cdot NH + NR_3' \\ \quad\Big| \qquad\qquad\qquad\quad\Big| \qquad\qquad \rightarrow \\ {}^{-}OCO \cdot NH \qquad R \cdot CH \cdot CO \cdot {}^{+}NR_3' \end{array}$$

$$\begin{array}{l} R \cdot CH \cdot CO \cdot NH \\ \quad\Big| \qquad\qquad\quad\Big| \qquad\qquad\qquad + CO_2 + NR_3', \\ {}^{-}O \cdot CO \cdot NH \quad R \cdot CH \cdot CO \cdot {}^{+}NR_3' \end{array}$$

and the repetition of these steps should yield a polypeptide.

WIELAND's mechanism accounts for the formation of diketo-piperazines, an interesting feature of his suggestion. This product results, according to his scheme, from an interaction of two "activated" monomers involving both of their ends, namely,

$$\begin{array}{l} {}^{+}NR_3' \\ \quad\Big| \\ R \cdot CH \cdot CO \qquad {}^{-}O \cdot CO \cdot NH \qquad R \cdot CH \cdot CO \cdot O \cdot CO \cdot NH \\ \quad\Big| \qquad\quad + \qquad \Big| \qquad\qquad \rightarrow \quad \Big| \qquad\qquad\qquad\qquad \Big| \qquad + 2\,NR_3' \rightarrow \\ NH \cdot COO^{-} \qquad CO\!-\!CH \cdot R \qquad NH \cdot CO \cdot O \cdot CO \cdot CH \cdot R \\ \qquad\qquad\qquad\quad \Big| \\ \qquad\qquad\qquad\quad {}^{+}NR_3' \end{array}$$

$$\begin{array}{l} R \cdot CH \cdot CO \cdot NH \\ \quad\Big| \qquad\qquad\quad \Big| \qquad\qquad + 2CO_2 + 2NR_3'. \\ NH \cdot CO \cdot CH \cdot R \end{array}$$

For the sake of brevity, this reaction was written in one step, whilst it probably proceeds through several stages, first forming a linear anhydride which then decarboxylates and finally cyclises. It seems significant that diketopiperazine is formed on reacting tertiary amines with sarcosine NCA[1] and its yield may be as high as 35% [see ref. (3), p. 318]. It should be stressed, however, that this process involves an N-substituted NCA, whereas in the polymerisation of the non-N-substituted NCA's no substantial amounts of diketopiperazines, if any,

---

[1] It is suggested that this product is formed *only* in the presence of water. No reaction was observed in anhydrons systems. (Private communication of Prof. BAMFORD.)

were observed. This again might indicate different mechanisms of polymerisation for these two types of NCA's.

The separation of terminal charges required by WIELAND's mechanism is its most unsatisfactory feature, i. e. the open chain form,

$$R_3'N^+ \cdot CO \cdot CH(R) \cdot NH \cdot \{CO \cdot CH(R) \cdot NH\}_n \cdot CO \cdot CH(R) \cdot NH \cdot COO^-,$$

is most unlikely in solvents of low dielectric constant. The contact between the terminal charges should lead to cyclisation, as exemplified by the formation of diketopiperazines, and it might be that cyclic polypeptides are partially formed by this route. It is also improbable to expect an interaction of two "activated" monomers in the process of initiation. The interaction of a non-activated molecule with an "activated" one would be much more plausible, and such a step was indeed proposed by BALLARD and BAMFORD (43).

In their original suggestion, which was modified later, these workers visualised the formation of a "complex", or an "activated" monomer between NCA and the aprotic base, or its positive ion, if the initiator is an ion-pair. Such a complex conceptually resembles the one proposed by WIELAND. For example, the initiation by lithium chloride in dimethylformamide solution was represented by the equation (44),

$$
\begin{array}{c}
\text{R} \cdot \text{CH—CO} \\
\left| \right\rangle\text{O} + \text{Li}^+ \rightleftharpoons \\
\text{NH—CO}
\end{array}
\quad
\begin{array}{c}
\text{R} \cdot \text{CH—CO} \\
\left| \right\rangle\text{O} \rightleftharpoons \\
\text{NH—CO, Li}^+
\end{array}
\quad
\left.
\begin{array}{c}
\text{R} \cdot \text{CH—CO} \\
\left| \right\rangle\text{O} \\
^-\text{N-----CO, Li}^+
\end{array}
\right\} \text{H}^+
$$

and a similar species was proposed for the initiation by alkali salts of weak carboxylic acids (54). The activated NCA then reacts with a non-activated monomer, regenerates the initiator, and gives a dimeric carbamic acid, viz.

$$
\left.
\begin{array}{c}
\text{R} \cdot \text{CH—CO} \\
\left| \right\rangle\text{O} \\
^-\text{N-----CO, Li}^+
\end{array}
\right\} \text{H}^+ +
\begin{array}{c}
\text{R} \cdot \text{CH—CO} \\
\left| \right\rangle\text{O} \rightarrow \\
\text{NH—CO}
\end{array}
$$

$$
\begin{array}{c}
\text{R} \cdot \text{CH—CO} \\
\left| \right\rangle\text{O} \\
\text{N-----CO, Li}^+ \\
\left.
\begin{array}{c}
\text{R} \cdot \text{CH—CO} \\
\left| \right\rangle\text{O} \\
\text{NH—CO}
\end{array}
\right\} \ominus, \text{H}^+
\end{array}
\rightarrow
\begin{array}{c}
\text{R} \cdot \text{CH—CO} \\
\left| \right\rangle\text{O} \\
\text{N-----CO} \\
\text{R} \cdot \text{CH—CO} \\
\left| \right. \\
\text{NH} \cdot \text{COOH}
\end{array}
+ \text{Li}^+.
$$

Finally, the latter decarboxylates into a bifunctional, polymerisation-initiating species,

$$
\begin{array}{l}
\text{R} \cdot \text{CH—CO} \\
\quad | \qquad\quad \diagdown \text{O} \\
\quad | \qquad\quad \diagup \\
\text{NH—CO} \\
\quad | \\
\text{CO} \cdot \text{CH(R)} \cdot \text{NH}_2 \,.
\end{array}
$$

It is assumed that the labile proton, denoted symbolically by H^+, is attached to a base associated with the positive ion, e. g. to Cl^- if lithium chloride is the initiator or to carboxylate ion if the initiation results from the action of sodium salt of a weak carboxylic acid (54).

The initiation by tertiary amines is proposed to proceed similarly and to give (43),

$$
\begin{array}{c}
{}^+\text{NR}_3' \\
| \\
\text{R} \cdot \text{CH—C—OH} \\
\quad | \qquad\quad \diagdown \text{O} \,, \\
\quad | \qquad\quad \diagup \\
{}^-\text{N—CO}
\end{array}
$$

i. e. a tautomeric form of Wieland's zwitter-ion. However, the labile proton is assumed in this case to be attached to the "activated" monomer rather than to another molecule of NR_3' and, as previously proposed, the interaction with a non-activated NCA regenerates the initiating base and gives the dimer which decarboxylates into the difunctional amine, i. e.

$$
\begin{array}{ccc}
{}^+\text{NR}_3' & & \\
| & & \\
\text{R} \cdot \text{CH—C—OH} & \text{R} \cdot \text{CH—CO} & \\
\quad | \qquad\quad \diagdown \text{O} & \quad | \qquad\quad \diagdown \text{O} & + \text{NR}_3' \\
\quad | \qquad\quad \diagup & \quad | \qquad\quad \diagup & \\
\text{N—CO} & \text{N—CO} & \\
| & | & \\
\text{R} \cdot \text{CH—CO} & \text{CO} \cdot \text{CH(R)} \cdot \text{NH} \cdot \text{COOH} & \\
| & & \\
\text{NH—COO}^- & &
\end{array}
$$

$$
\begin{array}{cc}
& \text{R} \cdot \text{CH—CO} \\
\rightarrow & \quad | \qquad\quad \diagdown \text{O} \qquad + \text{NR}_3' \,. \\
& \quad | \qquad\quad \diagup \\
& \text{N—CO} \\
& | \\
& \text{CO} \cdot \text{CH(R)} \cdot \text{NH}_2 + \text{CO}_2
\end{array}
$$

Perhaps it was unfortunate that these basic ideas evolved from Bamford's studies of NCA polymerisation initiated by LiCl in dimethyl formamide — certainly not the most typical polymerising system initiated by an aprotic-base. The presence of LiCl in the solution raised doubts in the minds of other investigators as to whether some specific effects, caused by the relatively high concentration of electrolyte, are not imposing a peculiar character on this polymerisation. Moreover, the LiCl system drew Bamford's attention to the positive ion, rather than to the base, and led him to the above formalism. The essential features of his mechanism may be preserved, and presented in an even more convincing manner if one postulates initiation involving a transfer of proton from the monomer to the base instead of the addition of the base to the monomer. This, indeed, was proposed by BAMFORD and BLOCK (51) in their paper published in 1961. The initiation was represented by the equation,

$$\begin{array}{ccc}
\text{R} \cdot \text{CH——CO} & & \text{R} \cdot \text{CH——CO} \\
| \qquad\quad \diagdown & & | \qquad\quad \diagdown \\
\quad\qquad\qquad \text{O} + \text{base} \rightleftharpoons & & \qquad\qquad\quad \text{O} + \text{Base} \cdot \text{H}^+, \\
| \qquad\quad \diagup & & | \qquad\quad \diagup \\
\text{NH——CO} & & {}^-\text{N}\text{-----CO}
\end{array}$$

followed by the conventional steps of BAMFORD's mechanism, leading to the dimeric carbamates,

$$\begin{array}{c}
\text{R} \cdot \text{CH——CO} \\
| \qquad\qquad \diagdown \\
| \qquad\qquad\quad \text{O} \\
| \qquad\qquad \diagup \\
\text{N——CO} \\
| \\
\text{CO} \cdot \text{CH(R)} \cdot \text{NH} \cdot \text{COO}^-,
\end{array}$$

which neutralise the protonated base and thus regenerate the initiating proton-acceptor. Finally, the decarboxylation of carbamic acid yields the dimeric bifunctional species possessing an amine on one of its ends and the active oxazolidone ring on the other.

In fact, this interpretation became imperative when it was found that lithium perchlorate in dimethyl formamide does *not* initiate the polymerisation in systems for which lithium chloride is an efficient initiator. This proves that the reaction involves the negative ion, i. e. Cl^- or ClO_4^-, and *not* the positive Li^+ ion, and while Cl^- in dimethyl formamide is a sufficiently strong base capable of accepting a proton and initiating the process, the ClO_4^- apparently is not. Actually, one may question to what extent these salts are dissociated in dimethyl formamide. It is possible, therefore, that the reaction involves ion-pairs rather than free ions, and the Li^+, Cl^- ion-pair may be a more powerful proton acceptor than Li^+, ClO_4^-.

To prove their hypothesis, BAMFORD and BLOCK (51) applied the diagnostic test previously designed by GOLD and JEFFERSON (58) in their studies of hydrolysis of carboxylic anhydrides catalysed by tertiary bases. The technique employed involves the use of a series of tertiary bases having different relative abilities to associate with Lewis acids and to act as Brönsted bases. Pyridine, α-picoline and 2,6-lutidine form such a series. The ortho-methyl group in picoline, or two such groups in lutidine, enhance the proton acceptability of the ring N-atom, making 2,6-lutidine the strongest Brönsted base and pyridine the weakest. On the other hand, the shielding of the N-atom by the adjacent methyl groups inhibits the addition of 2,6-lutidine to the carbonyl group and therefore, for such an addition, pyridine is the most reactive while 2,6-lutidine is the least. In studies of GOLD and JEFFERSON (58) pyridine was the most powerful catalyst en-hancing the hydrolysis of anhy-drides. Hence, in that reaction the catalytic action was caused by the addition of tertiary base to the car-bonyl group of the anhydride. The reverse order of activities was ob-served by BAMFORD and BLOCK (57) (see Fig. 13) and this result conclu-sively proves that the initiation of polymerisation by aprotic bases arises from proton abstraction and not from the addition of the base to on NCA molecule.

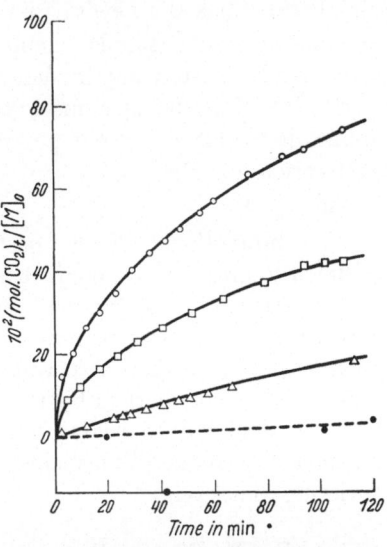

Fig. 13. Conversion-time curves for the polymeri-zation of N-carboxy-γ-ethyl-L-glutamate anhy-dride (0.2 mol l⁻¹) catalysed by pyridine and its ho-mologues (2.0 mol l⁻¹) in NN-dimethylformamide at 25°. △ pyridine, □ α-picoline, ○ 2,6-lutidine. The broken curve shows a comparative experiment with N-carboxysarcosine anhydride (0.2mol l⁻¹) and 2,6-lutidine (2.0 mol l⁻¹). [Reprinted from paper by C. H. BAMFORD and H. BLOCK: J. Chem. Soc. 4989 (1961) (Fig. 1)]

The concept of proton-transfer to the base leads to an important generalisation (18, 52), namely, this type of polymerisation may be in-itiated not only by aprotic bases but also by primary or secondary amines. The latter may then act in a dual way — as proton acceptors as well as proton donors. Hence, the initiation by primary or secondary base may lead to two simultaneous reactions competing for the base — one which irreversibly removes the base from the system,

$$R \cdot CH{-}CO \diagdown_{O} + HNR_1R_2 \leftrightarrows \quad R \cdot CH \cdot CO \cdot NR_1R_2 \quad \rightarrow \text{ polymer,}$$

and the second which is only catalysed by the base but does not affect its ultimate concentration, viz.

$$R \cdot CH{-}CO \diagdown_{O} + HNR_1R_2 \leftrightarrows \quad R \cdot CH{-}CO \diagup_{O} + H_2NR_1R_2^+ .$$

Therefore, two types of polymerisation ensue and this explains, e. g., the results of Blout and Karlson (49) (see p. 22) who found, for the reaction initiated by n-hexylamine in dioxane, a degree of polymerisation *higher* than expected on the basis of the "simple" amine-propagated mechanism.

The different behaviour of, e. g., di-isopropyl amine and n-hexyl amine now becomes explicable. The sterically unhindered n-hexyl amine mainly functions as a Lewis base which adds to the Lewis acid (the NCA) and initiates the "simple" amine-propagated polymerisation. On the other hand, the sterically hindered di-iso-propyl amine is inefficient as a Lewis base but, being a more powerful Brönsted base than n-hexyl amine, it initiates a rapid polymerisation resulting from the proton abstraction and the "activation" of the monomer. This is seen in Fig. 7 (see also p. 19).

The bifunctional dimer, postulated in Bamford's initiation, is assumed to grow by two distinct mechanisms. Its amine-terminated end may propagate in the "normal" fashion, but since this mode of growth is relatively slow, its contribution to the overall process is expected to be of *minor* significance. According to Bamford and Block (18, 51) the bulk of NCA polymerises by the mechanism involving an "activated" monomer and it is the latter mode which is assumed to be fast. The "monomer-activation" mechanism is analogous to the formation of the initiating dimer, viz.

The newly-formed activated monomer then continues the chain process, and the decarboxylation of the carbamic acid produces the previous type of polypeptide but one unit longer.

Increasing concentration of initiator has a dual effect on the rate of such polymerisation. It increases the stationary concentration of growing species — a trivial effect expected in nearly all polymerisation processes — but in addition it greatly enhances the rate of propagation of each growing polymer by increasing the concentration of the activated monomers. This situation is not encountered in conventional polymerisation processes where the rate of growth of each chain is independent of the initiator's concentration.

BAMFORD's mechanism offers an interesting possibility of coupling the polymeric molecules, i. e. the conventional reaction of the terminal amine of an n-mer with the cyclic group of an m-mer should produce an (n + m)-mer. This process, if it occurs, gives some poly-condensation character to this essentially poly-addition type of polymerisation. However, although such a condensation may rapidly increase the molecular weight of the produced polypeptide, it contributes relatively little to the CO_2 evolution or to the rate of monomer consumption. It should be interesting, therefore, to investigate the change in the molecular weight of the polypeptide in the last stages of the reaction. The only available information pertinent to this problem was reported by IDELSON and BLOUT (50) and it does not support the coupling hypothesis — the molecular weight of the polypeptide was found to remain constant as the conversion (i. e. the CO_2 evolution or the monomer consumption) increased from 25% to 100%. Investigation of this problem in other systems is highly desirable.

The coupling process cannot account for the high rate of the reaction since the number of growing chains is exceedingly small even at the onset of the reaction. Therefore, the statement of BALLARD and BAMFORD [ref. (43), p. 387], attributing the rapid rate of reaction to the coupling process, is probably erroneous.

The intramolecular coupling of the terminal amine with the cyclic end of the same polypeptide represents an interesting termination process (44, 54). Such a reaction results in the formation of cyclic polypeptides, and indeed, a hexameric cyclic polypeptide was isolated (44, 54) from the products of polymerisation initiated by some aprotic bases. The rate of such termination should depend on the molecular weight of the terminated polymer since the probability of ring closure is the greatest for a relatively low degree of polymerisation, say 4—8, and thereafter it decreases with increasing length of the chain. Hence, a substantial fraction of growing polymers should be terminated by cyclisation when their degree of polymerisation is low, whereas those which survived this critical stage should attain considerable size before being terminated — probably by some other processes. This feature of the termination may account for the large $\overline{M}_w/\overline{M}_n$ ratio of polypeptides produced in a polymerisation initiated by aprotic bases[1] and for the constant DP of the polymer produced in consecutive "after" polymerisations (see p. 25).

The most probable mode of termination of long-chain polymers may result from a wrong addition of an activated monomer, e. g.

$$
\begin{array}{ccc}
\text{R} \cdot \text{CH—CO} & \text{O} & \text{R} \cdot \text{CH—CO} \cdot \text{O} \cdot \text{CO} \\
\mid \quad \diagdown \text{O} + \text{CO} \quad \text{CO} \rightarrow & \mid \qquad\qquad \mid & \rightarrow \\
\text{\small www}\text{CO} \cdot \text{N—CO} & \underline{\text{N}}\text{——CHR} \qquad \text{\small www}\text{CO} \cdot \text{N——CO——N} \cdot \text{CH(R)COO}^{-}
\end{array}
$$

$$
\begin{array}{c}
\text{R} \cdot \text{CH—CO} \\
\mid \qquad \diagdown \\
\qquad\qquad \text{N} \cdot \text{CH(R)} \cdot \text{COO}^{-} + \text{CO}_2 , \\
\text{\small www}\text{CO} \cdot \text{N——CO}
\end{array}
$$

and not from a spontaneous termination such as cyclisation. The former type of termination, being first order with monomer, leaves the $\overline{DP}_w$ of the product independent of the initiator's concentration, whereas a spontaneous termination demands a decrease in $\overline{DP}_w$ on decreasing the initiator's concentration at a constant monomer concentration (i. e. on decreasing the concentration of the activated monomer).

It is also possible that the termination of the long chain polymers may be caused by some impurities present in the system since it has been frequently observed (private communication of Dr. A. Berger) that purification of the monomer and solvent leads to faster polymerisation.

The intramolecular coupling of the terminal amine with the oxazolidine ring attached to the other end of the polymeric chain involves the C-5 carbonylic group since the C-2 CO group appears to be much less reactive (see, e. g. p. 6). However, for steric reasons such a reaction is prevented in the initiating dimer, hence the amine of the dimer may react only with the 2-carbonyl group, viz.

$$
\begin{array}{ccc}
\text{R} \cdot \text{CH—CO (5)} & & \text{CH(R)} \cdot \text{COOH (or COO}^{-}) \\
\mid \quad \diagdown & & \mid \\
\qquad \text{O} & & \\
\text{N——CO (2)} & \rightarrow & \text{N——CO} \\
\mid \quad \diagdown & & \mid \quad \diagdown \\
\qquad \text{NH}_2 \text{ (or NH}^{-}) & & \qquad \text{NH} \\
\text{CO—CH(R)} & & \text{CO—CH(R) ,}
\end{array}
$$

giving a derivative of hydantoin-3-acetic acid. This isomerisation competes for the dimer with the propagation of polymerisation and

---

[1] The intermolecular coupling does not lead to a high ratio $\overline{M}_w/\overline{M}_n$. The high $\overline{M}_w/\overline{M}_n$ ratio arises from the presence of a substantial mole fraction of very low molecular weight product and not from a process forming a small fraction of high molecular weight polymer.

it reduces, therefore, the efficiency of the initiating process. Such description is preferred to that used by BAMFORD's group who classified the isomerisation of the dimer into hydantoin-3-acetic acid as a termination process.

It was found (43) that dilution of the polymerising solution increases the yield of the hydantoin-3-acetic acid and this result, predicted by BAMFORD's mechanism, provides additional evidence in its favour.

Two additional mechanisms of propagation should be considered. In the presence of strong bases such as sodium methoxide or sodium triphenyl methyl, the terminal amine may be converted, at least partially, into respective sodium amide. Hence, the following propagation may ensue under these conditions,

$$\text{\textasciitilde\textasciitilde\textasciitilde} NH^-, Na^+ + NCA \rightarrow \text{\textasciitilde\textasciitilde\textasciitilde} NH \cdot CO \cdot CH(R) \cdot NH \cdot COO^-, Na^+ .$$

The carbamate salts are stable and do not decarboxylate. Therefore, to account for the polymerisation one must assume either a proton transfer to carbamate, e. g. by reversing the process creating the amide or by providing protons from another source. Alternatively, one may postulate (58a) the activation of the NH by the adjacent carboxylate ion and represent the propagation by the equation,

$$\text{\textasciitilde\textasciitilde\textasciitilde} NH \cdot COO^- + NCA \rightarrow \quad \begin{array}{l} \text{\textasciitilde\textasciitilde\textasciitilde} NH + CO_2 \\ | \\ CO \cdot CHR \cdot NH \cdot COO^-. \end{array}$$

A similar mode of propagation was postulated by IDELSON and BLOUT (53) who assumed the formation of a linear anhydride as an intermediate and then its eventual decarboxylation, viz.

$$\text{\textasciitilde\textasciitilde\textasciitilde} NH \cdot COO^- + O\!\!\diagdown\!\!\begin{array}{c} CO{-}CH(R) \\ | \\ CO{-}NH \end{array} \rightarrow \begin{array}{c} \text{\textasciitilde\textasciitilde\textasciitilde} NH \cdot CO \cdot O \cdot CO \cdot CH(R) \\ | \\ NH \cdot COO^- \end{array} \rightarrow$$

$$\rightarrow \text{\textasciitilde\textasciitilde\textasciitilde} NH \cdot CO \cdot CH(R)NH \cdot COO^- + CO_2 .$$

The propagation involving the activated monomer-anion is possible only for the non-N-substituted NCA's — the N-substituted anhydrides cannot form such ions. Hence, if the polymerisation initiated by aprotic bases proceeds entirely by this route, no N-substituted NCA may be polymerised by these initiators [see however references (102), (103) and (104)]. Therefore, BAMFORD's group spared no effort to prove the validity of this statement and their elegant experiments deserve detailed discussion.

As shown in Fig. 7, sarcosine NCA, the N-substituted anhydride, polymerises faster than γ-ethyl-L-glutamate NCA if the polymerisation is initiated by n-hexyl amine in dimethyl formamide (*18, 52*). However, sarcosine NCA is found to be extremely unreactive when di-isopropyl amine is used for initiation of the reaction, while the polymerisation of γ-ethyl-L-glutamate becomes even faster under these conditions (*18, 52*) (see again Fig. 7). Steric hindrance inhibits the "simple" amine-addition initiation of di-isopropyl amine, and hence the slowness of sarcosine NCA polymerisation is not surprising. However, the high basicity of this initiator may result in the proton-transfer activation of the monomer, and this fast mode of polymerisation is operative for γ-ethyl-L-glutamate NCA but not for sarcosine NCA which lacks the necessary proton. Thus, the results shown in Fig. 7 are fully rationalised.

An even more convincing argument is provided by the studies (*46*), results of which are also shown in Fig. 14. Tri-n-butyl amine in dimethyl formamide induces only a slow polymerisation of L-proline NCA whereas the polymerisation of γ-ethyl-L-glutamate NCA is very fast under these conditions even if the concentration of the base is reduced by a factor of ten (*18, 46*). This is a striking observation since proline NCA is an extremely reactive monomer and polymerises very fast on addition of a primary amine. However, addition of 3-methyl hydantoin,

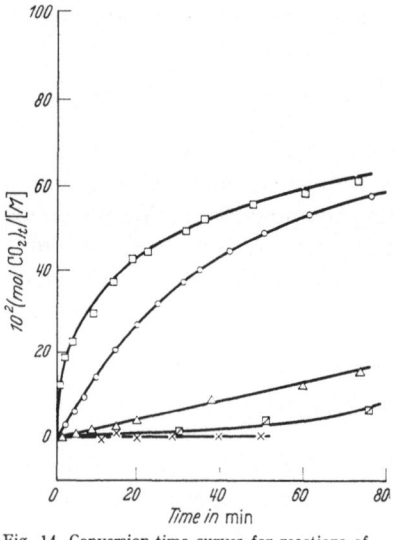

Fig. 14. Conversion-time curves for reactions of NCA's (*M*), with tri-n-butylamine (*B*) in N,N-dimethyl-formamide at 25° C. □ γ-Ethyl-L-glutamate NCA ($[M]_0 = 0.1$ mol l⁻¹, $[B] = 0.0126$ mol l⁻¹); ○ Sarcosine NCA ($[M]_0 = 0.332$ mol l⁻¹, $[B] = 0.042$ mol l⁻¹) + 3-methyl hydantoin (0.178 mol l⁻¹); △ L-Proline NCA ($[M]_0 = 0.291$ mol l⁻¹, $[B] = 0.129$ mol l⁻¹); ⊿ Sarcosine NCA ($[M]_0 = 0.289$ mol l⁻¹, $[B] = 0.262$ mol l⁻¹); × Sarcosine NCA ($[M]_0 = 0.323$ mol l⁻¹, $[B] = 0$ mol l⁻¹) + 3-methyl hydantoin (0.175 mol l⁻¹). [Reprinted from paper by C. H. Bamford and H. Block: Polyamino Acids, Polypeptides and Proteins, p. 65. Wisconsin University Press 1962 (Fig. 7)]

$$CH_2-CO$$
$$| \qquad \qquad \searrow N \cdot CH_3 ,$$
$$NH-CO \nearrow$$

to the investigated mixture substantially enhanced the polymerisation of L-proline NCA as may be seen in Fig. 14. Bamford therefore drew the obvious conclusion, namely, that 3-methyl hydantoin, which is expected to behave similarly to non-N-substituted NCA's, reacts with the base and produces the anion

$$CH_2-CO$$
$$| \qquad \qquad \searrow N \cdot CH_3 .$$
$$\underline{N}-CO \nearrow$$

This, in turn, interacts with the N-substituted NCA and gives the respective dimeric carbamic anion,

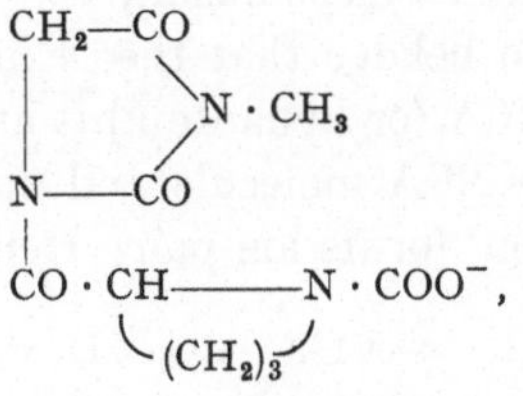

$$
\begin{array}{l}
\text{CH}_2\text{—CO} \\
\qquad\qquad\diagdown \\
\qquad\qquad\qquad \text{N} \cdot \text{CH}_3 \\
\qquad\qquad\diagup \\
\text{N}\text{——CO} \\
\mid \\
\text{CO} \cdot \text{CH}\text{———N} \cdot \text{COO}^-, \\
\qquad\diagdown_{(\text{CH}_2)_3}\diagup
\end{array}
$$

which becomes neutralised by the protonated base, viz.

$$
\text{Dimer. COO}^- + \text{Base} \cdot \text{H}^+ \;\rightarrow\; \text{Dimer. COOH} + \text{Base} \,.
$$

Decarboxylation of the acid then produces the amine group which continues the propagation. Of course, the hydantoin ring, being stable, cannot contribute to the growth and furthermore no "activated" monomer may be formed from proline-NCA. The argument was substantiated (46) by isolating and identifying the postulated intermediate, which was formed in the reaction of N-phenyl glycine NCA with a tertiary base and 3-methyl hydantoin. Since the substituted aniline is too weak a base to sustain the polymerisation, the reaction produced only a decarboxylated dimer,

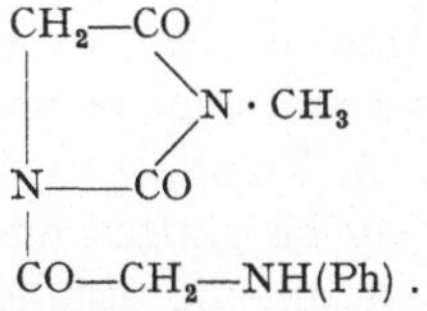

$$
\begin{array}{l}
\text{CH}_2\text{—CO} \\
\qquad\qquad\diagdown \\
\qquad\qquad\qquad \text{N} \cdot \text{CH}_3 \\
\qquad\qquad\diagup \\
\text{N}\text{——CO} \\
\mid \\
\text{CO—CH}_2\text{—NH(Ph)} \,.
\end{array}
$$

However, the exceptionally high capacity of 3-methyl hydantoin to initiate such a polymerisation [see ref. (46)] calls for some further explanations.

The results discussed above raise the following questions:

(1) Is the initiation of N-substituted NCA's polymerisation by aprotic bases prevented by the lack of intermediates having the structure of NCA anion, viz.

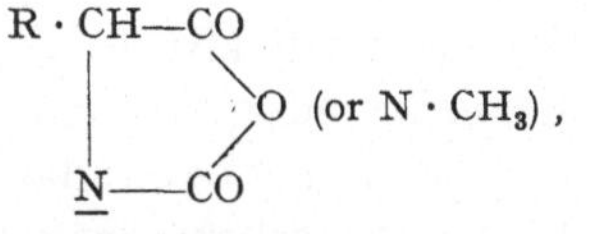

$$
\begin{array}{l}
\text{R} \cdot \text{CH—CO} \\
\qquad\qquad\diagdown \\
\qquad\qquad\qquad \text{O (or N} \cdot \text{CH}_3)\,, \\
\qquad\qquad\diagup \\
\text{N}\text{——CO}
\end{array}
$$

or is the decarboxylation of the resulting carbamic acid prevented by the lack of a suitable proton donor.

(2) Is the fast polymerisation of the *non*-N-substituted NCA's, which leads to high molecular weight polymer, the result of propagation

involving the "activated" monomer attacking the terminal ring, or is it caused by a mechanism involving the terminal $\sim$NH$^-$ or $\sim$NH $\cdot$ COO$^-$ group (BLOUT's or WALKER's mechanism).

There is no reason to believe that the $\overline{N}$ ion (or the corresponding ion-pair) derived from NCA (or hydantoin) is an exceptional base, more efficient in attacking an NCA molecule and opening its ring than any other sufficiently basic ion[1] (or its ion-pair). Hence, the reactions such as

$$NCA + {}^-OCH_3 \rightarrow CH_3O \cdot CO \cdot CH(R) \cdot NH \cdot COO^-$$

or

$$NCA + {}^-C(Ph)_3 \rightarrow C(Ph)_3 \cdot CO \cdot CH(R) \cdot NH \cdot COO^-$$

should also be efficient in initiating polymerisation if the growth results from BLOUT's type of propagation. However, since under normal conditions leading to high molecular weight polymer the concentration of NCA is much greater than that of the base, and since proton-transfers are faster than the base additions, the anions derived from NCA should be virtually the *only* bases capable of initiating polymerisation, the originally added base being converted into its conjugated acid. Therefore, the evidence presented by GOODMAN and ARNON (55), who did not find the fragments of initiator in the polymer, is to some extent misleading. This observation does not disprove the intrinsic possibility of initiation by the addition of CH$_3$O$^-$ ions but merely shows that these ions virtually are absent under conditions of their experiments, as well as under conditions prevailing in conventional polymerisation systems. In fact, unpublished results of Dr. A. ZILKHA (private communication) showed that polypeptide chains may be grafted on cellulose containing –ONa groups. This implies that an anionic residue may be incorporated into polypeptide chain.

The basic problem is raised by the second question, namely, which mechanism of propagation operates in the polymerisation giving the *high molecular weight* polypeptide. The writer proposes, therefore, the following experiment which may resolve the present controversy. Let us prepare a radio-active salt of carbamic acid, say,

$$CH_3O \cdot CO \cdot C^{14}H(R) \cdot NH \cdot COO^-, Na^+$$

and add to its solution the investigated NCA. If BLOUT's type of propagation operates, a fast reaction should ensue immediately and yield a high molecular weight polymer containing one radio-active group per chain. On the other hand, if BAMFORD's scheme operates, one may anticipate no polymerisation or polymerisation which starts slowly and then

---

[1] According to Prof. BAMFORD 3-methyl-hydantoin anion is exceptional in its ability to initiate the polymerisation of *N-substituted* NCA's in the presence of tertiary amines. This observation calls for some additional explanation.

accelerates, the process being initiated by the sequence,

(1) $CH_3 \cdot O \cdot CO \cdot C^{14}H(R) \cdot NH \cdot COO^- + NCA \rightleftharpoons$

$\rightleftharpoons CH_3O \cdot CO \cdot C^{14}H(R)NH \cdot COOH +$ activated NCA^- ion.

(2) The carbamic acid decarboxylates and thus the above-given equilibrium would be displaced to the right.

(3) The activated monomer should then initiate the polymerisation leading to the high molecular polymer containing *no radio*-activity.

The radio-active ester of amino acid may initiate an "amine" propagated polymerisation, but this reaction should be of little importance if the conversion is limited, say, to 20%. Hence, the presence of activity in *high* molecular weight polymer, and lack of induction period should prove BLOUT's type of propagation whilst its absence and induction period supports BAMFORD's type of propagation involving attack of an activated monomer on a polymer possessing a diketo-oxazolidone ring on its end.

At the time when the writer received the galleys of this review the full report of GOODMAN and ARNON (*102*) appeared in the Journal of the American Chemical Society. In their most interesting paper they described an experiment in which the polymerisation of γ-benzyl-L-glutamate NCA was successfully initiated by sodium N-benzyl-carbamate. This experiment comes closely to the suggestion proposed by this writer but still it is inconclusive.

Goodman and Arnon's goal was to disprove the thesis that the polymerisation initiated by a strong base is propagated by the free amine (the "simple" amine-propagated mechanism) or by amide-ion, viz.

$\text{\textasciitilde}CO \cdot CH(R) \cdot NH^- + NCA \rightarrow \text{\textasciitilde}CO \cdot CH(R) \cdot NH \cdot CO \cdot CH(R) \cdot NH \cdot COO^- \rightarrow$

$\rightarrow \text{\textasciitilde}CO \cdot CH(R) \cdot NH \cdot CO \cdot CH(R) \cdot NH^- + CO_2$, etc.

Their experiments proved that these modes of propagation do not operate in the strong base initiated polymerisation. The addition of methyl iodide to polymerising mixture had *no* effect upon the process. Since the iodide methylates the free amine, a polymerisation propagated by such a group should be inhibited. This indeed is the case if methyl iodide is added to a mixture being polymerised by the "simple" amine propagated mechanism [experiments conducted by J. HUTCHINSON and quoted in ref.(*102*)], but no effect whatsoever is observed in a polymerisation initiated by a strong base.

Again, the addition of methyl methacrylate had no effect upon the rate of polymerisation or upon the structure of the resulting poly-peptide. Since, amide groups initiate the polymerisation of methyl methacrylate the absence of such a reaction proves the absence of the amide groups.

This writer does not doubt this evidence. He accepts the conclusion that the "fast" polymerisation initiated by a strong base is *not* propagated by a free amine or by an amide, and indeed these are *not* the modes of propagation claimed by Bamford and Block (*18*). The issue is whether the polymerisation is propagated by the carbamate, as proposed by Blout and endorsed by Goodman and Arnon, or is it propagated through the "active" NCA anion, as proposed by Bamford and Block. In fact, Goodman and Arnon provided a very elegant proof of initiation of the polymerisation by such an anion, i. e., by an "activated" NCA. They agree, therefore, with Bamford and Block that the resulting polymer possesses a terminal oxazolidine ring. Since the reaction of an "activated" monomer with NCA, leading to a dimer possessing a terminal ring, is confirmed by Goodman and Arnon, the reaction of an "activated" monomer with a terminal oxazolidine ring should be equally acceptable — and this is precisely the propagation mechanism proposed by Bamford and Block (*18*). The evidence presented by Goodman and Arnon (*102*) does not refute *this* mode of propagation.

The experiment described by Goodman and Arnon (*102*) in which sodium N-benzylcarbamate was the initiator has to be modified to prove the occurrence of Blout's type of initiation. One should label with C^{14} the benzyl group of the carbamate (e. g., $Ph \cdot C^{14}H_2 \cdot NH \cdot COONa$) and then to demonstrate the retention of the activity by the polymer. On the other hand, the loss of activity would demonstrate that

$$Ph \cdot CH_2 \cdot NH \cdot COO^-, Na^+ + NCA \rightleftarrows Ph \cdot CH_2 \cdot NH \cdot COOH + O \underset{CO-CHR}{\overset{CO-\bar{N}, Na^+}{\diagup}}$$

The decarboxylation of carbamic acid shifts the equilibrium to the right, and the "activated" NCA would then initiate Bamford's type of propagation. This is essentially the test proposed in the earlier paragraph by the writer, and its result should unequivocally discriminate between the two rival mechanisms of propagation.

It should be remarked, at this juncture, that Bamford's experiments utilising 3-methyl hydantoin as a co-catalyst do not shed light on the problem of propagation. They have shown, merely, that a proton is required for the "simple" amine-propagated polymerisation and, according to his scheme, any proton donor may suffice to produce this result (see however the footnote on p. 44). For example, if his mechanism is correct, sarcosine NCA should not be polymerised by sodium methoxide in a rigorously aprotic medium[1], but polymerisation should ensue on

---

[1] Notice that Goodman and Arnon (*55*) report polymerisation of sarcosine NCA initiated by sodium methoxide. It is doubtful, however, whether their system was rigorously aprotic. In fact, Dr. Goodman agrees that his experiments were not performed under rigorously aprotic conditions. However, he feels also that sarcosine would be polymerised by sodium methoxide even if the medium were rigorously aprotic.

addition of methanol to the non-reacting mixture. Of course, such an experiment cannot be performed with tri-n-butyl amine, since the steric hindrance will prevent its action as a Lewis base.

The recent proposal of GOODMAN and ARNON (55) is pertinent to these problems. They suggest the abstraction of a C–H proton from the N-substituted NCA's as a possible mode of initiation of polymerisation, e. g.,

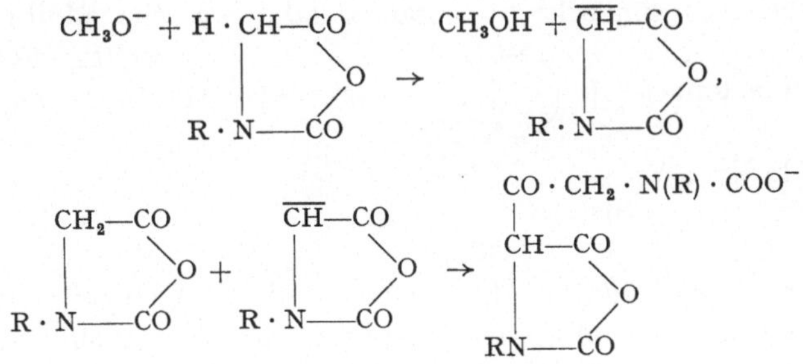

$$CH_3O^- + H \cdot CH\text{—}CO \qquad CH_3OH + \overline{CH}\text{—}CO$$

If Blout's mechanism is correct — the carbamate would continue then the propagation. Alternatively, neutralisation of the carbamate ion by methanol and the decarboxylation leads then to

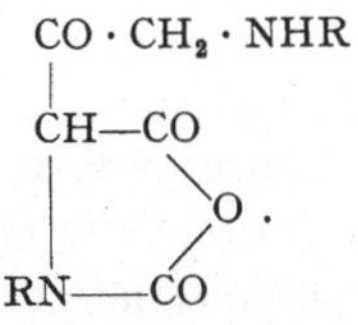

$$CO \cdot CH_2 \cdot NHR$$
$$CH\text{—}CO$$
$$RN\text{—}CO$$
$$O .$$

This dimer may then grow through "simple" amine-propagated steps, since the growth from the ring end involving the C activated monomer will not produce a polypeptide, especially if the reaction proceeds on the usually most reactive CO(5) group. The proposed formation of the NCA anion having a negative charge on the C atom was tested by adding to an excess of sodium methoxide the L-proline NCA and observing its racemisation.

More evidence has been accumulated [see e. g. ref. (58)] to show that the polymerisation yielding high molecular weight polypeptides proceeds in two steps — initial self-accelerated reaction followed by an apparently first order reaction. It seems that the growing species slowly reach their stationary concentration and in this period the reaction appears to be auto-catalytic. In the terms of BAMFORD's mechanism this behaviour is easily explained by postulating slow initiation and rapid propagation. The initiation results from an attack of an activated monomer on a non-activated NCA. The propagation results from a similar reaction involving the species

$$\text{\textasciitilde\textasciitilde CO} \cdot \text{N} \text{------} \text{CO}$$

$$\text{CH(R)} \text{---} \text{CO}$$

which may be much more reactive than the free NCA due to an acti-
vating effect of the adjacent carbonylic group. In this respect the NCA
system behaves similarly to lactames (60), e. g. the addition of an

activated lactame's salt $\left(\begin{array}{c}\text{-CO}\\ |\\ \text{-NH}^-, \text{A}^+\end{array}\right)$ proceeds faster to $\left(\begin{array}{c}\text{Ph}\cdot\text{CO}\cdot\text{N}\\ |\\ \text{OC}\end{array}\right)$

$\left(\text{or } \begin{array}{c}\text{\textasciitilde\textasciitilde CO}\cdot\text{N}\\ |\\ \text{OC}\end{array}\right)$ than to $\begin{array}{c}\text{NH}\\ |\\ \text{OC}\end{array}$.

It is not surprising that the rate of aprotic NCA polymerisation and
the molecular weight of the resulting product are extremely susceptible
to the nature of the base, counter-ion, added electrolyte, solvent etc.
These factors affect the equilibrium,

$$\text{NCA} + \text{base} \rightleftharpoons \text{activated NCA}^- + \text{base} \cdot \text{H}^+ ,$$

and therefore the rate of polymerisation if it indeed proceeds by the
activated NCA mechanism. Moreover, if the termination involves a
"wrong" addition of the activated NCA$^-$ then the nature of the solvent
and of the counter-ion (hence also of Base $\cdot$ H$^+$ if it is an ion) should
affect the ratio k_p/k_t. This in turn affects both the molecular weight of
the product and the rate of polymerisation (by affecting the stationary
concentration of growing ends). It is interesting to recall at this stage
the peculiar fact, namely, the polymerisation of some NCA's does not
take place in tri-n-butyl amine as a solvent, but it proceeds in dimethyl
formamide in the presence of a small amount of the amine. One may
wonder whether the equilibrium,

$$\text{NCA} + \text{Bu}_3\text{N} \rightleftharpoons \text{activated NCA}^- + \text{Bu}_3\text{N}^+ ,$$

is not displaced too much to the right in Bu$_3$N being a solvent, leaving
too low concentration of free NCA to allow an efficient initiation.

The "activated" monomer propagation is only applicable to the non-
N-substituted NCA's. The polymerisation of the N-substituted NCA
iniated by strong bases must proceed by a different mechanism. The
initiation may result from a process proposed by Goodman and Arnon
(55), although such an initiation does not seem plausible for tertiary
amines. The proton-donating impurities facilitate this reaction as shown
by Bamford et al. (46). Furthermore, the Blout or Walker type of
propagation, or the closely related Wieland's propagation, might operate

in this case. Let us recall, e. g., the high yield of diketopiperazine produced under these conditions from sarcosine NCA [see ref. (3)]. Notice, however, that if an aprotic base initiates a genuine polymerisation of N-substituted NCA's, which is not propagated by an amine, or an amide, then there must exist a new mode of propagation, e. g. such as proposed by Blout.

Finally, we should consider some objections which were levelled against the activated monomer hypothesis and which were recorded in the literature. There is some confusion about the significance of the bifunctional dimer proposed by BAMFORD. For example, BLOUT (61) remarked that the fast polymerisation cannot be ascribed to the free amine end. This is correct, but the activated monomer mechanism operates on the ring-terminated end and this is the supposedly fast reaction which yields a high molecular weight product. BLOUT emphasises also that in the polymerisation initiated, e. g., by triphenyl methyl sodium the free amine end cannot exist but should be converted into a respective sodium amide. Since the contribution of free amine propagation is insignificant when compared with that of the activated monomer propagation, its conversion into another species is of little importance. However, under conditions which lead to a high molecular weight polymer, the free amine is most probably unaffected by the base. Since the concentration of NCA exceeds that of the free amine, and the former is a stronger acid than the latter, the equilibrium in the reaction,

$$\text{~~~NH}^-, \text{Na}^+ + \text{NCA} \leftrightarrows \text{~~~NH}_2 + \text{activated NCA}^-, \text{Na}^+,$$

is displaced far to the right.

Another misconception arises from the statement that a strong base, e. g., $C(Ph)_3^-$, Na^+ will be converted into an extremely weak acid, i. e. Ph_3CH, and will not provide a proton for the decarboxylation of carbamate ions. This again is true, but the proton need not be provided by the conjugated acid — it is given by the non-activated NCA which is converted in this process into an activated NCA. This reaction cannot take place with the N-substituted NCA's and therefore the activated monomer mechanism is not operative for these monomers.

The most serious objection is the lack of experimental evidence for the existence of the ring compound on the end of the polymer (62). Apart from the general difficulty of detecting and identifying end-groups of high molecular weight polymers, an additional problem arises in this system. The reaction *does* involve a termination step and this must destroy the terminal ring group responsible for the growth.

In conclusion, the complexity of NCA polymerisation arises from the numerous modes of reaction of these monomers. Collating all the available evidence, we find that the nucleophiles and bases may interact

with NCA in two fashions: through addition in which the monomer behaves as a Lewis acid or by proton abstraction in which it acts as a weak
proton donating acid. The growth takes place either by the addition of
non-activated monomer to a free amine group, or perhaps to $-NH \cdot COO^-$,
or by the addition of an activated monomer to an oxazolidine ring formed
on the end of a growing polymer. The latter mode is probably fast and
might give the high molecular weight polymer. The variety of termination steps add to the complex character of this polymerisation.

11. Physical factors in the kinetics of NCA polymerisation

The complexity of NCA polymerisation is often magnified by some
phenomena of a physical rather than a chemical nature, and those
caused by the state of aggregation of the polymerising system or by the
shape of polymeric molecule may play a profound role in determining
the course of these reactions.

Many attempts were made to describe the structure of proteins
in terms of helices stabilised by intramolecular hydrogen bonds. The
pioneering work of Astbury and his school (66) indicated that molecules
of proteins, especially of wool, may exist in an extended or in a folded
configuration, the latter being considered as a helix. These ideas were
developed further and applied to poly-amino acids by Elliott, Bamford
and their associates (67, 68, 69). However, the precise structure of
proteins and poly-amino acids was firmly established by the extensive
studies of Pauling and Corey (64). Their X-ray work proved that the
molecules of fibrous keratine form an α-helix, i. e. an array composed of
13-membered rings, closed by intramolecular hydrogen bonds, forming
spirals with a pitch of 5.4 A and containing about 3.6 amino-acid residues
per turn of the spiral. The consecutive amino-acid residues are axially
transposed by 1.50 A. The α-helix appears to be the most stable helical
structure — it imposes no strain on the polymeric chain, leaves no
empty space in the centre and satisfies the linearity of hydrogen bonds.
It rapidly assumed a dominant role in subsequent studies when it was
demonstrated that the molecules of many other proteins and polyamino-
acids acquire such a shape. Nevertheless, other helical bondings, anticipated by Pauling, must not be renounced. For example, Luzzati and
his colleagues (65) reported transitions of α-helices into more elongated
3_{10} helices, in which the consecutive amino-acid residues are displaced
by 1.95 A.

The folded form of proteins, reported by Astbury, was identified
with the α-helix, the extended one described in terms of intermolecularly
hydrogen-bonded array of chains was known in the earlier literature as the

β-form. In films made from synthetic polyamino acids the α and β forms could be transformed one into the other. The folded form is established if films are casted from m-cresol solutions, whereas in the films casted from formic acid solution the polymer takes the extended β form. These observations show that the solvent affects the shape of the polymeric molecule.

The helical structure of polyamino acids may persist in solution and this fact, which had far-reaching consequences for further developments, was recognised by DOTY and his colleagues (70). In a short note, published in 1954, they reported striking differences in the hydrodynamic behaviour of synthetic, high-molecular weight poly-γ-benzyl-L-glutamate, caused by changing the solvent. The studies of viscosity, light-scattering, IR spectra etc. revealed that these polymeric molecules behave in dioxane or chloroform as rigid, rod-like particles, whereas in dichloroacetic acid they acquire the conventional shape of random-coils characterising the solution of ordinary polymers. The rod-like molecules were recognised as α-helices. In dichloroacetic acid the helical structure is unstable, since the intermolecular bonds between the amino-acid residues and the powerfully solvating $CHCl_2 \cdot COOH$ replace the intramolecular hydrogen bonds of the helix, and therefore the chain acquires the randomly-coiled shape.

Further work led to realisation that, even in a non-destructive solvent, a perfect helical conformation cannot be maintained along the entire length of a high-molecular weight polymer. Hence, longer or shorter helical stretches have to be linked by disorientated, randomly-coiled chains — the degree of disorientation being determined by temperature, nature of the solvent, of the polymer and perhaps by its molecular weight (71). The existence of the orientated and disorientated portions of the chain poses the interesting problem of helix-random coil transition (72). While the atoms in the coiled portion of the polymeric molecule perform an endless Brownian dance, their position becomes fixed in the helical stretches which form one-dimensional crystals. The spontaneous "crystallisation" of the disorientated chains presents a fascinating aspect of the physical-chemistry of polyamino-acids, and this phenomenon has been observed only in one more system, namely in polynucleic acids and polynucleotides. The question arises, therefore, what is the smallest number of residues which permits spontaneous formation of a helix.

To answer this question, GOODMAN and his team synthesised by classic techniques a series of polypeptides of γ-methyl-L-glutamate ranging from 3 to 11 units all composed of the same enanthiomorphic form of the amino-acid residues (73). The specific optical rotation of a polymeric molecule involving asymmetrical centres is enhanced by its helical structure (74), and therefore the shape of an oligomer may be recognised from the degree of optical rotation of its solution. In view of this, Goodman's group examined the optical rotation of their oligomers in a variety of solvents (75), their results being shown in Fig. 15. It may be seen that the specific optical rotation in dichloro-acetic acid *decreased*

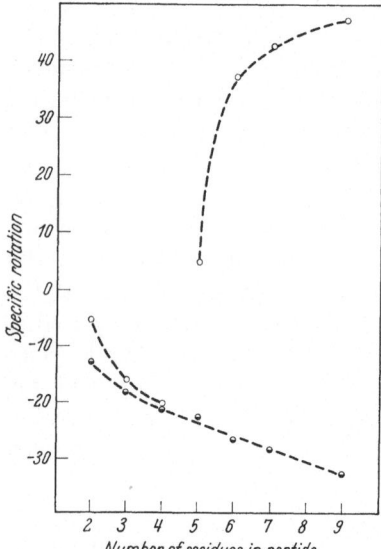

with increasing degree of polymerisation (see the full circles in Fig. 15). However, in dioxane the decrease of the rotation was observed only for the first 3 oligomers ($n = 2, 3$ and 4), whereas an abrupt increase in the rotation was shown in the solutions of pentamer and even higher values were found for the solutions of hexa-, hepta- and nonamers.

An increase in the specific rotation of low-molecular weight polypeptides may be caused also by their intermolecular association (75). Such an effect should be concentration dependent, and indeed in dioxane the specific rotations of the penta- and hexamer were found to increase with

Fig. 15. Optical activity of peptide derivatives as a function of solvent and number of residues. All rotations were measured in dioxane (open circles) and dichloroacetic acid (half-filled circles) at 2% concentration except those for the hepta- and nonapeptides in dioxane solution. These rotations were measured on a 1.43% and 0.22% solution, respectively. [Reprinted from J. Am. Chem. Soc. **81**, 5507 (1959). Communications to the Editor (Fig. 1)]

their concentration, and ultracentrifugation studies (77) proved indeed that these oligomers are associated. However, the specific optical rotations of the dimer, trimer and tetramer, as well as of the hepta- and nonamer were concentration independent, and hence the abnormally high specific rotation of the hepta- and nonamers indicates that they are probably helical.

To avoid any complications caused by the intermolecular association, Goodman (78) reinvestigated the optical rotation of the peptides in dimethyl formamide, since in this medium the specific rotation is independent of concentration. From the latter study it was concluded that at 25° C the spontaneous helix formation of poly-γ-methyl-L-glutamate in dimethyl formamide is occurring at the critical range of 7—9 units. Extension of these studies (73b, 79) led to a better understanding of temperature and solvent effects upon the helix-coil transition of oligomeric polypeptides.

Alternatively, the problems of helix-coil transition may be investigated by determining the UV absorption of a polypeptide in the

190 mμ region since the extinction coefficient of this band is markedly increased by the formation of a helix (80). This criterion was adopted in the studies of GOODMAN and LISTOWSKY (81).

Recently, studies of the conformation of oligomers were extended to peptides derived from β-methyl-L-aspartates. Their synthesis ($n = 2$ up to 14) was described by GOODMAN and BOARDMAN (82), and later the specific rotations of their solutions in dimethyl formamide, dichloro-acetic acid and in chloroform were determined (83). The oligomers exist in a random-coil form in the first two solvents, but helices become stable in chloroform for $n = 11$ and 14. These peptides are unusual since their L-amino-acid residues produce a left-hand helix (84, 85) whereas most of the investigated polyamino acids "crystallise" as a right-hand helix (86).

A most interesting study of helix stability has been completed recently by LIQUORI (87). His computing programme permits the calculation of the total repulsive energy of a helix arising from the interaction between non-bonded atoms as a function of two or three angles. For example, the inclinations of the CH(R)–NH–CO and NH–CO–CH(R) planes in respect to the helix axis are employed if R=H or CH_3, and an additional angle is introduced to describe the position of group R, if R contains 2 or more carbon atoms. The calculations assume the conventional bond lengths and bond angles, as well as the planarity of the –NH–CO group. The results show that the energy, represented as a function of the variable angles, has exceptionally deep and sharp minima and hence their values, and therefore the pitch of the helix, may be accurately determined. The treatment shows unambiguously whether the right-handed or left-handed helix is preferred, and the agreement with the observations is remarkable.

Similar calculations were previously performed for helical, isotactic polymeric hydrocarbons possessing the $-CH_2-CR_1R_2-$ recurring units (88). In this case the variable angles describe the inclinations of the $-CH_2-C(R_1R_2)-CH_2$ and $C(R_1R_2)-CH_2-C(R_1R_2)$ planes. The minima are not as sharp as those calculated for the polyamino-acids, and the symmetry of these chains makes, of course, the right-handed and left-handed helix equally probable[1].

Conversion of a randomly-coiled growing polymer into a helix affects the surroundings of its active end and, therefore, such a transformation is expected to modify the rate constant of propagation. This most interesting situation was visualised by DOTY and LUNDBERG (36), and a model, taken from a paper by WEINGARTEN (33) and shown in Fig. 16, illustrates how a helical structure might influence the monomer addition.

---

[1] Actually, in this chain the distinction between the right-handed and left-handed helix is purely formal.

A molecule of NCA is assumed to be hydrogen-bonded to the terminal active amine of a growing helix and simultaneously to the NH group of a residue separated by three amino-acid units from the polymer's end. If the resulting orientation of the NCA is beneficial for its addition, then

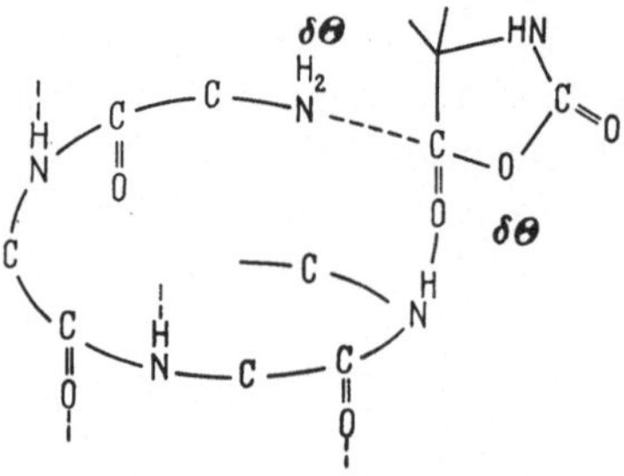

Fig. 16. [Reprinted from paper by H. Weingarten: J. Am. Chem. Soc. **80**, 352 (1958)]

the destruction of the helix, or even replacement of the N–H hydrogen-bonded amino-acid residue by its enanthiomorph, destroys the favourable configuration and, therefore, this should slow down the polymerisation. One may also adapt the model of Weingarten to account for a fast growth of a helix terminated by an oxazolidine ring. For example, it could be argued that a hydrogen bond between the terminal ring and a suitable unit of the helix activates the oxazolidine residue and enhances, therefore, the propagation.

Leaving apart the details of this mechanism, which still need verification, we may now examine the available evidence supporting the hypothesis of helical growth. The first indication of such a reaction, which has been discussed previously (see p. 18), came from the thorough kinetic study (17) of γ-benzyl-L-glutamate polymerisation initiated by n-hexyl amine in dioxane. The initiation of this process was found to be very rapid ($k_{in} \approx 0.2$ l/mole sec), and the propagation was described in terms of two second-order rate constants: k_{2a} — referring to the first slow stage of the process, and k_{2b} — characterising the second fast period. The rate constants were calculated from the slopes of the respective straight lines (see Fig. 6) *assuming* the concentration of growing ends to be given by the initial concentration of n-hexyl amine. In spite of a 10-fold change in the monomer-to-initiator ratio (from 4 to 40), k_{2a} and k_{2b} were found to be reasonably constant. Addition of monomer to a polymerised solution led to further polymerisation, its rate constant being the same as that of the preceding fast reaction. For example, when the polymerisation, initiated at $M/I = 20$, was followed to 98 % of conversion, k_{2b} was determined as 0.035 l/mole sec; subsequent addition of monomer and solvent, to give an M/I ratio of 12, re-initiated the reaction which proceeded in a perfectly pseudo-unimolecular fashion, its second-order rate constant k_2 being 0.038 l/mole sec. It appears, therefore, that neither the concentration of growing polymer nor the rate constant had not change any more after the polymer reached a critical size of 8—11 units. On the basis of this evidence, Lundberg and Doty concluded that the slow reaction results from the growth of a randomly-coiled, low-molecular weight peptide, while the fast polymerisation is caused by the propagation of a helix. In agreement with this hypothesis the polymerisation initiated at M/I ratio of 4 showed only the first slow stage, $k_{2a} = 0.007$ l/mole sec; obviously, no helices can be formed at such a low DP and therefore the fast reaction could not possible ensue.

Two additional facts support the hypothesis of helical growth:

(1) In dimethyl formamide, only the first slow stage of the reaction was observed. Helices are less stable in this solvent and therefore their spontaneous formation could be inhibited.

(2) The IR studies of IDELSON and BLOUT (53) demonstrated that in dioxane the initially-formed polymer possessed a random-coiled structure (described by the authors as the β-form). As the reaction proceeded a new material appeared, which was identified through its IR spectrum as the α-helix. This stage of the process coincided with the onset of a fast reaction. By using a deuterated n-hexyl amine as the initiator a labelled β-peptide was prepared. This was used in turn to initiate further polymerisation which yielded an α-peptide. The isolation of the latter and its analysis proved that the α-peptide contained the expected percentage of deuterium, and hence the β-polymer had to be the precursor of α-peptide.

Although the phenomenon of "two-stage polymerisations" is well established and it had been observed by other investigators (53, 50, 59), it seems that this polymerisation does not involve two pseudo-first order reactions, as suggested by LUNDBERG and DOTY, but its kinetics arises from a relatively slow approach to the stationary state imposing an autocatalytic character on the overall process. Moreover, it is likely that other factors, perhaps jointly with a helical growth, contribute to its acceleration. The following observations lead to this conclusion:

(1) The growth of random coils is uniform and the onset of helix stability should occur rather abruptly at some critical size of the polymer (71). Hence, the absence of termination should lead to narrow molecular weight distribution (26), whereas the experimental results indicated a large $\overline{M}_w/\overline{M}_n$ ratio ranging from 3 to 8. This implies that only a small fraction of the initially-formed polymers yields a high-molecular weight material and, therefore, k_{2b} was incorrectly calculated and its constancy appears to be fortuitous[1].

LUNDBERG and DOTY tried to explain the high $\overline{M}_w/\overline{M}_n$ ratio by assuming that a small fraction of polymers, namely those which reached the critical size, consumed in their rapid growth all the available monomer. Hence, they starved the remaining chains and thus prevented them from attaining the critical length. This idea was formalised mathematically by COOMBES and KATCHALSKI (89), who derived the expression for

---

[1] Unpublished results of Dr. BLOUT indicate that the high molecular weight polypeptide formed by sodium methoxide initiation does not show abnormally broad molecular weight distribution.

the molecular weight distribution of the resulting polymer. Unfortunately, their calculations do not agree with the experimental data.

(2) The kinetics of γ-benzyl-D,L-glutamate NCA polymerisation (*17*) resembles that of the L-isomer. Again, two stages are discerned in the

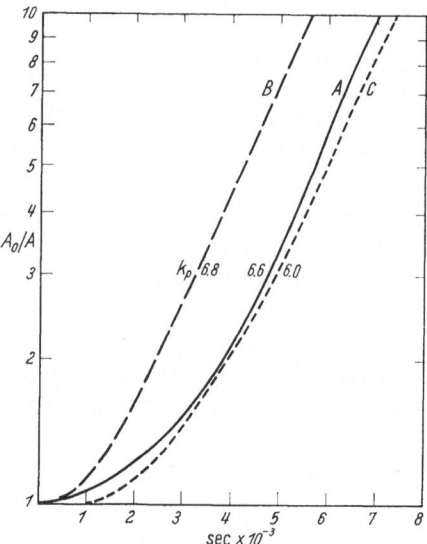

process, although the onset of the fast growth appears later than in the polymerisation of the L-isomer. LUNDBERG and DOTY suggested that to stabilise a helical structure the D,L polymer has to be longer than an all L-polymer. However, it was shown by HUGGINS (*90*) that the presence of a wrong enanthiomorph in a sequence of four amino-acid residues makes the helical structure improbable and calculations of LIQUORI fully confirm this prediction. Hence, the self-acceleration of the racemic polymerisation should be attributed to other factors than helical growth.

Fig. 17. Curve A ———, plot of anhydride consumption (A_0/A) measured by CO_2 evolution method as a function of time for an A/I 400 sodium methoxide initiated polymerization of γ-benzyl-L-glutamate-N-carboxyanhydride in dioxane solution (4 g/100 ml). Curve B — — —, the rate curve obtained upon the addition of an equal quantity of L-NCA to a completed L-polymerization. Curve C - - - -, the rate curve obtained upon the addition of an equal quantity of D-NCA to a completed L-polymerization. The k_p values are shown in a horizontal line with the symbol. [Reprinted from paper by IDELSON and BLOUT: J. Am. Chem. Soc. **80**, 2387 (1958) (Fig. 2)]

(3) The self-acceleration of γ-benzyl-L-glutamate NCA polymerisation initiated by sodium methoxide was reported by IDELSON and BLOUT (*50*). This reaction produces a high-molecular weight polymer and acceleration, although less pronounced, is still observed when fresh monomer is added to a polymerised solution (see Fig. 17). It is significant that the straight portions of curves A and B, shown in Fig. 17, are parallel. Apparently, the same stationary state eventually is attained in both reactions.

The above arguments do not disprove the reality of helical polymerisation: they show only that the kinetic evidence may not be sound. A stronger proof for such a growth is provided by the study of the effects exerted by the presence of one enanthiomorph on the propagation of the other. This approach was also explored by LUNDBERG and DOTY, and their results are given in Table 8. The racemic monomer, as well as the mixture of both isomers, polymerises more slowly than the pure enanthiomorph. The identical rates found for D,L and D + L mixture prove that the observed effect is not caused by impurities. The reduction of the rate is small for $M/I = 4$ or for the slow stage of the process when $M/I = 20$, but in its fast stage it amounts to a factor of 2. These findings apparently demonstrate the preference of the growing end for its own enanthiomorph and such preference is enhanced in peptides of higher degree of polymerisation and hence the latter observation supports the idea of helical growth.

For a stereo-selective helical growth one might expect the racemic mixture to grow twice as slowly as the pure isomer — simply on account of decreased concentration of each isomer. This suggestion was examined by LUNDBERG and DOTY who added a small amounts of D-isomer to an

Table 8. *Propagation constants of isomeric and racemic γ-benzyl glutamate NCA. Initiated by n-hexyl amine in dioxane at 25° C*

Monomer	$M/I = 4$	$M/I = 20$	
		$k_{2a} \times 10^3$	$k_{2b} \times 10^3$
	$k \times 10^3$ 1/mole sec	1/mole sec	
L	7.4	5.6	32
D	7.1	7.4	34
D,L	5.1	4.5	13
D + L	5.5	4.2	15—17

From LUNDBERG and DOTY: J. Amer. Chem. Soc. **79**, 3961 (1957).

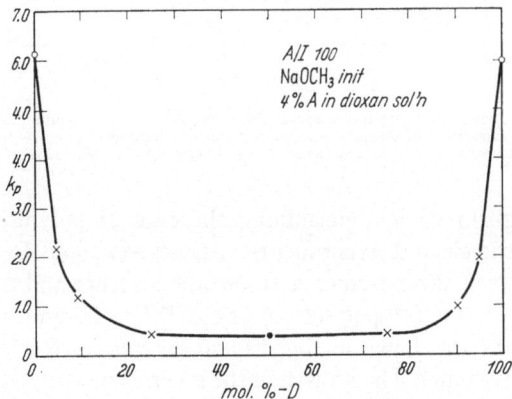

Fig. 18. The polymerization constants k_p, for sodium methoxide initiated polymerizations of γ-benzyl-L-glutamate-N-carboxyanhydride in dioxane solution (4 g/100 ml) as a function of mole per cent. D-anhydride isomer: ×, polymers made with DL-NCA and optically active NCA; O, polymers made with mixtures of pure D- and pure L-NCA. [Reprinted from paper by IDELSON and BLOUT: J. Am. Chem. Soc. **80**, 2387 (1958) (Fig. 3)]

L-monomer. The observed effect was, however, even greater than anticipated. In fact, IDELSON and BLOUT (50), who repeated these experiments in sodium methoxide initiated polymerisation that yields a high-molecular weight polypeptide, found enormous effects which are shown in Fig. 18. Addition of even 5% of D-isomer to an L-monomer reduced the polymerisation rate by a factor of 3, and the 50:50 mixture polymerised 17-times more slowly than either isomer. The co-operative effect of this phenomenon is therefore indisputable. The polymerisation is not 100% stereo-selective; however, its rate is determined not only by the last unit but also by some other segments, say, the fourth one from the end (see Fig. 16). To allow for a fast reaction both units must possess the same configuration and form a regular helix, and therefore incorporation of a D-unit into an all-L-polymer inhibits the polymerisation of *both* isomers. The fact that the effect was greatly magnified when high-molecular polymer was formed is another strong indication of a helical growth. A further proof for participation of penultimate segments in the growth,

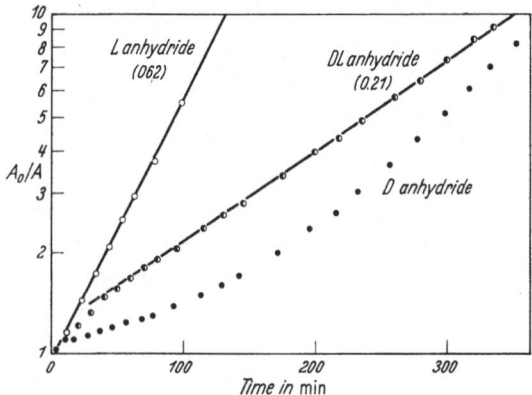

Fig. 19. Initiation by aliquote of preformed polymer (DP = 20 of polybenzyl-L-glutamate on L (○), DL (◑), and D (●), carboxy-anhydrides of γ-benzyl-glutamate in dimethylformamide; $[A_0] = 0.058$; $[I] = 0.0048$. [Reprinted from Lundberg and Doty: J. Am. Chem. Soc. **79**, 3961 (1957) (Fig. 10)]

and their importance in determining the rate of polymerisation, was furnished by the elegant experiments of Lundberg and Doty (17). The polymerisations of the L-isomer, L,D-racemic mixture and D-isomer were initiated by performed L-polymer of about $DP = 20$ and uniform in its size. The results are shown in Fig. 19 and demonstrate a "normal" fast growth of the L-isomer. The growth in the racemic mixture started slowly, and then decelerated, indicating a dilatorious effect of a wrong monomer on the growth by either NCA. In solution of D-isomer the initial growth was very slow but the polymerisation accelerated as D-segments were added to the L-helix. It appears that on addition of a sufficient number of D units to an L-polymer a new helical spiral is formed. Thus, the end of the resulting block polymer (D preceding L) behaves in the same way in respect to D monomer as the previous all-L polymer behaved in respect to an L monomer. This conclusion is verified by studies of changes in the optical rotation of the polymerising solutions. These are shown in Fig. 20. The increase and then decrease in the optical rotation when the polymerisation was investigated in solution of D-isomer is particularly striking. It indicates that a helix with an opposite sense of direction is formed when a sufficient number of D-units are added to an L-helix.

In conclusion, the reality of helical growth, which was proposed by LUNDBERG and DOTY, seems to be proved beyond any doubt. In fact, these phenomena may play an important role in other polymerising systems. SZWARC (91) suggested that the results of WILLIAMS et al. (92), who studied anionic polymerisation of styrene in different solvents, may be explained in terms of a helical growth, and a similar suggestion was made by HAM (93) to account for peculiarities of some radical polymerisations. Well-documented studies of anionic polymerisation of methyl methacrylate (94) lead to the conclusion that random-coiled polymers, as well as helices, may participate in this reaction. This subject has been fully reviewed in a recent publication (95).

The state of aggregation of the polymerising system represents another important factor which may affect the kinetics of polymerisation. It is well known (96,97) that many radical polymerisations are enhanced by increase in the viscosity of the polymerising system, and this pheno-

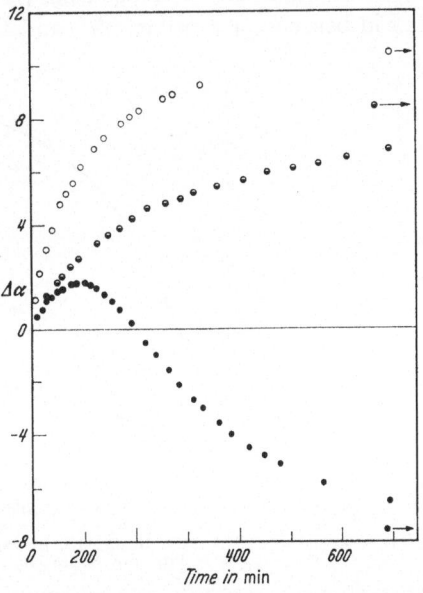

Fig. 20. The change in optical rotation of L (○), DL (◑), and D (●), carboxy-anhydride of γ-benzyl-glutamate during polymerization initiated by aliquots of an L-polymer; $[A_0] = 0.076$; $[I] = 0.0034$; polarimeter tube length 20 cm. [Reprinted from LUNDBERG and DOTY: J. Am. Chem. Soc. 79, 3961 (1957) (Fig. 11)]

menon was explained by a decrease in the rate of termination step which may become diffusion-controlled. In fact, the effect of viscosity should be observed at any stage of radical polymerisation, and this problem has been discussed recently by BENSON and NORTH (98, 99). Of course, this type of acceleration cannot be observed when the growth involves living polymers and therefore such an explanation does not apply to polymerisation of NCA, particularly since no termination resulting from active end-active end interaction takes place in these processes.

In some systems the polymer may precipitate in the course of the reaction and this again greatly affects the kinetics of polymerisation, e. g., in a radical polymerisation the precipitation may lead to the formation of "trapped" radicals. Moreover, separation into two phases affects the concentration of monomer around the growing centres and this may either speed up the propagation, if the polymer is a better solvent for the monomer than the original medium, whereas an inhibition is expected in the reverse case. These phenomena are particularly important in processes involving living polymers, or long-lived polymers, and therefore they should be observed in NCA polymerisation. In fact, in several NCA polymerisations a marked acceleration of the rate was noticed when the systems became heterogeneous (*100*). These processes were investigated in benzene or nitrobenzene, and the precipitated polymer appear to

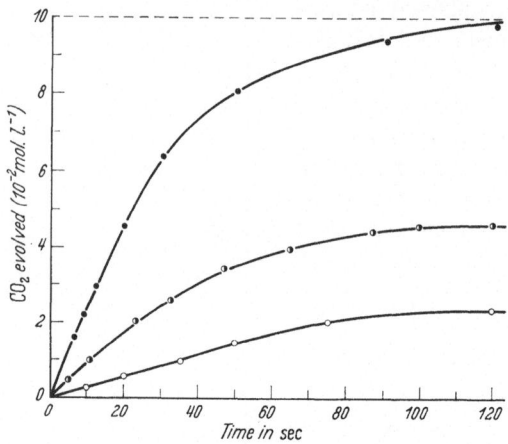

Fig. 21. Polymerization of DL-phenylalanine NCA initiated by: sarcosine dimethylamide (○); polysarcosine dimethylamide, $n = 5$ (◑) and $n = 10$ (●). - - - -, theoretical yield. Nitrobenzene solution; temperature 25° C; $[M]_0 = 0.10$ mol l⁻¹, $[X] = 8.6 \times 10^{-3}$ mol l⁻¹. [Reprinted from D. G. H. Ballard and C. H. Bamford: Proc. Roy. Soc. A **236**, 384 (1956) (Fig. 1)]

solvate the monomer better than the original solvents. Hence, the local increase in monomer's concentration seems to account for the observed phenomenon. The acceleration was also observed in polymerisation of sarcosine NCA. Since the resulting polymer possesses no N–H bonds, it cannot form a helix. Therefore, at least in this case, the acceleration cannot be accounted for by helical growth.

Although the heterogeneity of the system complicates the kinetics of polymerisation, and the partition of the monomer between the solvent and the precipitated polymer may accelerate the reaction, it seems that these observations do not disprove the reality of helical growth. The existence of this phenomenon is indicated most convincingly by the startling effect of D-isomer on the growth of L-polymer or vice versa.

A peculiar and intriguing effect of polysarcosine, which was used as an initiator of block polymerisation, was observed by Ballard and Bamford (*101*). Their results are shown in Fig. 21 and indicate that the initial rates of D,L-phenyl alanine NCA polymerisation are affected by the degree of polymerisation of the initiator. For a constant monomer concentration and a constant concentration of the initiating amine the rate significantly increases as the degree of polymerisation of poly-sarcosine rises from 1 to 10 [1]. These findings are summarised in Fig. 22 which shows that the initial rate of polymerisation approaches a constant value for $n \geq 20$.

The following facts should be noticed: (1) The polymerisation produces a block polymer — polysarcosine linked to the newly-formed

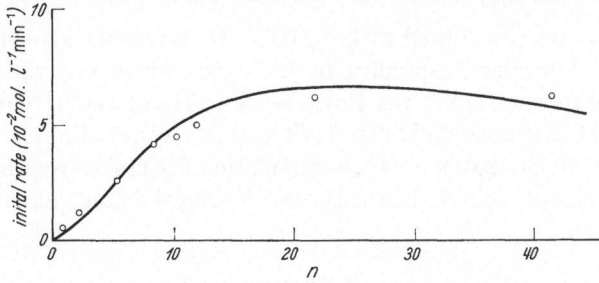

Fig. 22. Polymerization of DL-phenylalanine NCA initiated by polysarcosine dimethylamides with different degrees of polymerization. Nitrobenzene solution; temperature 15° C; $[M]_0 = 0.100$ mol l^{-1}, $[X] = 5.4 \times$ $\times 10^{-3}$ mol l^{-1}; ——, theoretical curve. [Reprinted from D. G. H. BALLARD and C. H. BAMFORD: Proc. Roy. Soc. A **236**, 384 (1956) (Fig. 2)]

peptide. (2) The effect was observed in the polymerisation of four different NCA's all initiated by polysarcosine. (3) The solvent plays an important role in the process. The reaction is greatly accelerated in chloroform and nitrobenzene and much less in dimethyl formamide.

Table 9. $[M]_0 = 0.105$ mole l^{-1}; nitrobenzene solution; 15° C

Initiator	Concentration (10^{-3} mole l^{-1})	Initial Rate (10^{-2} mole l^{-1} min^{-1})
Polysarcosine dimethylamide, $n = 20$	5	5.3
N-acetyl polysarcosine dimethylamide, $n = 20$.	5	0
N-acetyl polysarcosine dimethylamide, $n = 20$ + phenylalanine dimethylamide . . .	each 5	0.4
Phenylalanine dimethylamide	10	0.13

From BALLARD and BAMFORD: ref. (101).

(4) As shown in Table 9 acetylation of polysarcosine destroys the effect and the addition of such a polymer to phenylalanine dimethyl amide only slightly accelerates the process. (5) No effect is observed when sarcosine NCA is block-polymerised in poly-D,L-phenylalanine.

BALLARD and BAMFORD propose an interesting explanation for this phenomenon, to which they refer as "chain effect". The molecules of NCA supposedly are preferentially adsorbed to the polysarcosine chain and hence they are kept in the vicinity of the growing end. This increase in

---

[1] These effects were never observed in polymerisation of γ-benzyl-L-glutamate NCA (private communication by Prof. BLOUT).

"monomer's supply" leads to the fast growth. Their argument is strengthened by the observation that the addition of "dead" (acetylated) polysarcosine inhibits the polymerisation. This, they argue, removes the monomer which becomes bonded to an useless polymer.

The "chain effect" deserves further study. It is strange that it appears only in polysarcosine, since BAMFORD's explanation suggests that a similar mechanism may also operate for other polypeptiees.

The writer is indebted to Prof. C. E. H. BAWN for his helpful comments and for his hospitality in Liverpool where the writer held a Visiting Professorship of the Royal Society. He wishes to thank also to Prof. C. H. BAMFORD and to Dr. H. BLOCK as well as to Prof. E. R. BLOUT and Prof. M. GOODMAN for their stimulating discussions and for reading the manuscript. Finally he thanks to The Royal Society for the Visiting Professorship.

References

1. LEUCHS, H.: Ber. **39**, 857 (1906).
2. (a) LEUCHS, H., and W. MANASSE: Ber. **40**, 3235 (1907).
   (b) —, and W. GEIGER: Ber. **41**, 1721 (1908).
3. KATCHALSKI, E., and M. SELA: Advances in Protein Chem. **13**, 249 (1958).
4. BARTLETT, P. D., and R. H. JONES: J. Am. Chem. Soc. **79**, 2153 (1957).
5. MILLER, E., I. FANKUCHEN, and H. MARK: J. Appl. Phys. **20**, 531 (1949).
6. See, e. g. M. SELA, and E. KATCHALSKI: J. Am. Chem. Soc. **76**, 129 (1956).
7. DIECKMANN, W., and F. BREEST: Ber. **39**, 3052 (1906).
8. KOPPLE, K. D.: J. Am. Chem. Soc. **79**, 6442 (1957).
9. HEYNS, K., and R. BROCKMANN: Z. Naturforsch. **9**b, 21 (1954).
10. FUCHS, F.: Ber. **55**, 2943 (1922).
11. WESSELY, F.: Z. physiol. Chem. **146**, 72 (1925).
12. —, and F. SIGMUND: Z. physiol. Chem. **157**, 91 (1926).
13. HANBY, W. E., S. G. WALEY, and J. WATSON: J. Chem. Soc. **1950**, 3009.
14. WALEY, S. G., and J. WATSON: Proc. Roy. Soc. A **199**, 499 (1949).
15. FESSLER, J. H., and A. G. OGSTON: Trans. Faraday Soc. **47**, 667 (1951).
16. POPE, M. T., T. J. WEAKLEY, and R. J. P. WILLIAMS: J. Chem. Soc. **1959**, 3442.
17. LUNDBERG, R. D., and P. DOTY: J. Am. Chem. Soc. **79**, 3961 (1957).
18. BAMFORD, C. H., and H. BLOCK: In: Polyamino Acids, Polypeptides and Proteins, p. 65. Wisconsin University Press 1962.
19. BAILEY, J. L.: J. Chem. Soc. **1950**, 3461.
20. BALLARD, D. G. H., and C. H. BAMFORD: Proc. Roy. Soc. A **223**, 495 (1954).
21. HEYNS, K., H. SCHULTZE, and R. BROCKMANN: Ann. **611**, 33 (1958).
22. —, and H. SCHULTZE: Ann. **611**, 40 (1958).
23. SELA, M., and A. BERGER: J. Am. Chem. Soc. **77**, 1893 (1955).
24. KOPPLE, K. D.: J. Am. Chem. Soc. **79**, 662 (1957).
25. HANBY, W. E., S. G. WALEY, and J. WATSON: J. Chem. Soc. **1950**, 3239.
26. MITCHELL, J. C., A. E. WOODWARD, and P. DOTY: J. Am. Chem. Soc. **79**, 3955 (1957).
27. WESSELY, F., and M. JOHN: Z. physiol. Chem. **170**, 38 (1927).
28. WOODWARD, R. B., and C. H. SCHRAMM: J. Am. Chem. Soc. **69**, 1551 (1947).

28a. SLUYTERMAN, L. A., and A. LABRUYERE: Rec. trav. chim. **73**, 347 (1954).

29. COLEMAN, D., and A. C. FARTHING: J. Chem. Soc. **1950**, 3218.

30. BREITENBACH, J. W., and K. ALLINGER: Monatsh. Chem. **84**, 1103 (1953).

31. FRANKEL, M., and E. KATCHALSKI: J. Am. Chem. Soc. **65**, 1670 (1943).

32. FAURHOLT, C.: J. chim. phys. **22**, 1 (1925).

33. WEINGARTEN, H.: J. Am. Chem. Soc. **80**, 352 (1958).

34. SHALITIN, Y.: Ph. D. Thesis, Hebrew University (1958).

35. BLOUT, E. R., and A. ASADOURIAN: J. Am. Chem. Soc. **78**, 955 (1956).

36. DOTY, P., and R. D. LUNDBERG: J. Am. Chem. Soc. **78**, 4810 (1956).

37. KATCHALSKI, E., Y. SHALITIN, and M. GEHATIA: J. Am. Chem. Soc. **77**, 1925 (1955).

38. BALLARD, D. G. H., and C. H. BAMFORD: J. Am. Chem. Soc. **79**, 2336 (1957).

39. DOTY, P., and R. D. LUNDBERG: J. Am. Chem. Soc. **79**, 2338 (1957).

40. BALLARD, D. G. H., C. H. BAMFORD, and A. ELLIOTT: Makromol. Chem. **35**, 222 (1960).

41. BILEK, L., J. DERKOSCH, H. MICHL, and F. WESSELY: Monatschr. Chem. **84**, 717 (1953).

42. COLEMAN, D.: J. Chem. Soc. **1950**, 3222.

43. BALLARD, D. G. H., and C. H. BAMFORD: J. Chem. Soc. **1956**, 381.

44. — —, and F. J. WEYMOUTH: Proc. Roy. Soc. **227** A, 155 (1955).

45. KURTZ, J., G. D. FASMAN, A. BERGER, and E. KATCHALSKI: J. Am. Chem. Soc. **80**, 393 (1958).

46. BAMFORD, C. H., H. BLOCK, and A. C. P. PUGH: J. Chem. Soc. **1961**, 2057.

47. BLOUT, E. R., R. H. KARLSON, P. DOTY, and B. HARGITAY: J. Am. Chem. Soc. **76**, 4492 (1954).

48. DOTY, P., A. M. HOLTZER, J. H. BRADBURY, and E. R. BLOUT: J. Am. Chem. Soc. **76**, 4493 (1954).

49. BLOUT, E. R., and R. H. KARLSON: J. Am. Chem. Soc. **78**, 941 (1956).

50. IDELSON, M., and E. R. BLOUT: J. Am. Chem. Soc. **80**, 2387 (1958).

51. BAMFORD, C. H., and H. BLOCK: J. Chem. Soc. **1961**, 4989.

52. — — J. Chem. Soc. **1961**, 4992.

52a. PATCHORNIK, A., and Y. SHALITIN: Anal. Chemistry **33**, 1887 (1961).

53. IDELSON, M., and E. R. BLOUT: J. Am. Chem. Soc. **79**, 3948 (1957).

54. BALLARD, D. G. H., and C. H. BAMFORD: Special Publication of Chem. Soc. No. 2, p. 25 (1955).

55. GOODMAN, M., and U. ARNON: Biopolymers. **1**, 500 (1963).

56. WIELAND, T.: Angew. Chem. **63**, 7 (1951).

57. — Angew. Chem. **66**, 507 (1954).

58. GOLD, V., and E. G. JEFFERSON: J. Chem. Soc. **1953**, 1409 and 1416.

58a. WALKER, E. E.: Proceeding International Colloquium on Macromolecules. Amsterdam (1949), p. 381.

59. PEGGION, E., A. COSANI, A. M. MATTUCCI, and E. SCOFFONE: Biopolymers. **2**, 69 (1964).

60. CHAMPETIER, G., and H. SEKIGUCHI: J. Polymer Sci. **48**, 309 (1960) contains the previous references. See also for a review: M. SZWARC: Progress in Chemical Kinetics, vol. **4**, p. 226. Oxford: Pergamon Press 1964.

61. BLOUT, E. R., and M. IDELSON: Polyamino Acids, Polypeptides and Proteins, p. 79. The University of Wisconsin Press 1962.

62. KATCHALSKI, E., and M. SELA: ref. 3, p. 62.

63. Kopple, K. D., and J. J. Katz: J. Am. Chem. Soc. 78, 6199 (1956); J. Org. Chem. 27, 1062 (1962).

64. Pauling, L., R. B. Corey, and H. L. Branson: Proc. Nat. Acad. Sci. 37, 205 and 235 (1951).

65. Luzzati, V., M. Cesari, G. Spach, F. Marson, and J. M. Vincent: Polyamino Acids etc. p. 121. The Wisconsin University Press 1962.

66. (a) Astbury, W. T., and A. Idveet: Phil. Trans. Roy. Soc. A 230, 75 (1931).
    (b) —, and H. J. Woods: Phil. Trans. Roy. Soc. A 232, 333 (1933).

67. Ambrose, E. J., and A. Elliott: Proc. Roy. Soc. 205 A, 47 (1951).

68. Bamford, C. H., W. E. Hanby, and F. Happey: Proc. Roy. Soc. 206 A, 407 (1951).

69. Elliott, A.: Proc. Roy. Soc. 221 A, 104 (1953).

70. Doty, P., A. M. Holtzer, J. H. Bradbury, and E. R. Blout: J. Am. Chem. Soc. 76, 4493 (1954).

71. Schellman, C. G., and J. A. Schellman: Compt. rend. trav. lab., Carlsberg, Ser. chim. 30, 465 (1958).

72. Zimm, B. H., and J. K. Bragg: J. Chem. Phys. 28, 1246 (1958); 31, 526 (1959).

73. (a) Goodman, M., and K. C. Stueben: J. Am. Chem. Soc. 81, 3980 (1959).
    (b) —, E. E. Schmitt, and D. A. Yphantis: J. Am. Chem. Soc. 84, 1283 (1962).

74. (a) Doty, P., and J. T. Yang: J. Am. Chem. Soc. 78, 498 (1956); 79, 761 (1957).
    (b) See also W. Moffitt, and J. T. Yang: Proc. Nat. Acad. Sci. 42, 596 (1956).

75. Goodman, M., and E. E. Schmitt: J. Am. Chem. Soc. 81, 5507 (1959).

76. Doty, P., A. M. Holtzer, J. H. Bradbury, and E. R. Blout: J. Am. Chem. Soc. 76, 4493 (1954).

77. Yphantis, D. A.: Ann. N. Y. Acad. Sci. 88, 586 (1960).

78. Goodman, M., E. E. Schmitt, and D. A. Yphantis: J. Am. Chem. Soc. 82, 3483 (1960).

79. — — — J. Am. Chem. Soc. 84, 1288 (1962).

80. e. g. K. Rosenheck, and P. Doty: Proc. Nat. Acad. Sci. 47, 1775 (1961).

81. Goodman, M., and I. Listowsky: J. Am. Chem. Soc. 84, 3770 (1962).

82. —, and F. Boardman: J. Am. Chem. Soc. 85, 2483 (1963).

83. — —, and I. Listowsky: J. Am. Chem. Soc. 85, 2491 (1963).

84. Karlson, R. H., K. S. Norland, G. D. Fasman, and E. R. Blout: J. Am. Chem. Soc. 82, 2268 (1960).

85. Elliott, A., E. M. Bradbury, A. R. Downie, and W. E. Hanby: Proc. Roy. Soc. A 259, 110 (1960).

86. See e. g., E. R. Blout: In: Optical Rotatory Dispersion; chapter 17. McGraw-Hill 1961.

87. Liquori, A. M., et al.: In the course of publication.

88. Santis, P. de, E. Giglio, A. M. Liquori, and A. Ripamonti: J. Polymer Sci. 1 A, 1383 (1963).

89. Coombes, J. D., and E. Katchalski: J. Am. Chem. Soc. 82, 5280 (1960).

90. Huggins, M. L.: J. Am. Chem. Soc. 74, 3963 (1952).

91. Szwarc, M.: Chem. and Ind. p. 1589 (1958).

92. Williams, J. L. R., T. M. Laakso, and W. J. Dulmage: J. Org. Chem. 23, 638 (1958).

93. HAM, G. E.: J. Polymer Sci. **40**, 569 (1959).
 94. GLUSKER, D. L., I. LYSLOFF, and E. STILES: J. Polymer Sci. **49**, 315 (1961).
 95. SZWARC, M.: In: The Progress in Chemical Kinetics, Vol. II, p. 228. Oxford: Pergamon Press 1964.
 96. NORRISH, R. G. W., and R. R. SMITH: Nature **150**, 336 (1942).
 97. TROMSDORFF, E., H. KÖHLE, and P. LAGALLY: Makromolek. Chem. **1**, 169 (1947).
 98. BENSON, S.W., and A. M. NORTH: J. Am. Chem. Soc. **81**, 1339 (1959).
 99. NORTH, A. M., and G. A. REED: Trans. Faraday Soc. **57**, 859 (1961).
100. BALLARD, D. G. H., and C. H. BAMFORD: J. Chem. Soc. **1959**, 1039.
101. — — Proc. Roy. Soc. A **236**, 384 (1956).
102. GOODMAN, M., and U. ARNON: J. Amer. Chem. Soc. **86**, 3384 (1964).
103. FASMEN, G. D., and E. R. BLOUT: Biopolymers **1**, 3 (1963).
104. — — Biopolymers **1**, 99 (1963).

Adv. Polymer Sci., Vol. 4, pp. 66—110 (1965)

Polymerization Initiated by Lithium and its Compounds[1]

By

S. BYWATER

Applied Chemistry Division,
National Research Council,
Ottawa, Canada

With 9 Figures

Table of Contents

I. Introduction

Although the polymerization of olefins by heat and acid catalysts has been known since about 1840, polymerization by alkali metals and their alkyls was discovered more recently. Initiation by alkali metals was described by MATTHEWS and STRANGE (63), HARRIES (42), and

---

[1] Issued as N. R. C. No. 8044.

SCHLENK (*91*) in the period 1910—1914. The simple alkyls of lithium and sodium were first isolated by SCHLENK (*92*) in 1917 and polymerization initiated by these compounds was investigated intensively by ZIEGLER in subsequent years. In a review paper in 1936, ZIEGLER (*127*) put forward mechanisms for this type of polymerization which are surprisingly close to those accepted today. The mechanism of metal-alkyl initiated polymerization was formulated as:

$$R\text{--Met.} \xrightarrow{+M} R . M_1\text{--Met.} \xrightarrow{+M} R . M_2\text{--Met.} \ldots$$

whereas with alkali metal initiation it was suggested that it adds to both ends of a diene molecule to form an intermediate of the type

$$Met.CH_2\text{--}CH{=}CH\text{--}CH_2 . Met.$$

which subsequently adds monomer at both ends of the chain. ZIEGLER recognized that it is necessary to consider only two reaction steps, the initiation step (addition of the reagent to the first monomer molecule) and the subsequent chain propagation steps. The rate of initiation was shown to be strongly dependent on the nature of the alkyl or aryl group in the initiating compound. Thus the red colour of the triphenyl-methyl compound was shown to persist during a polymerization of butadiene in ether, whereas with butyllithium the initiator had disappeared early in the reaction. Some attempts were also made to determine the effect of solvent and alkali metal on the 1,2/1,4 ratio in polybutadiene by analysis of low molecular weight polymers. The results were misleading and it was not at that time realized that lithium compounds in non-polar solvents tend to give a highly specific product.

The discovery that lithium and its alkyls produce a highly cis-1,4 polyisoprene in hydrocarbon solvents (*103*) has led to a renewed interest in metal and metal alkyl initiated polymerization. About the same time SZWARC (*109*) postulated an electron transfer mechanism for the initiation of polymerization by sodium naphthalene in ether solvents. This was extended to lithium metal catalysis by TOBOLSKY (*80*) and OVERBERGER (*83*) and subsequently generalized to cover all alkali metal initiation.

$$e^- + M \rightarrow M^- \qquad (1) *$$
$$M^- + M \rightarrow {}^\cdot M . M^- \qquad (2)$$
$$e^- + {}^\cdot M . M^- \rightarrow {}^- M . M^- \qquad (3)$$
$$2 M^- \rightarrow {}^- M . M^- \qquad (4)$$
$$2 {}^\cdot M_n^- \rightarrow {}^- (M)_{2n}^- \qquad (5) .$$

The initial electron transfer produces a radical-ion (M^-) having an extra electron in its lowest anti-bonding π-orbital. The relative importance of subsequent steps has been discussed by SZWARC (*108*). The radical ends disappear rapidly by dimerization in most cases and the

---

* Counter-ions omitted for simplicity.

propagation proceeds at both ends by an anionic mechanism. The generally accepted mechanism for lithium alkyl initiated polymerization is still essentially that suggested by Ziegler with minor modifications.

Although our knowledge of the detailed processes involved in this type of polymerization has increased greatly in the last few years, a number of problems remain unsolved and this review can only be considered as a progress report.

II. Properties of lithium compounds
a) Lithium alkyls

The properties of lithium metal are well known, but the properties of its alkyls have until recently received much less attention. The lowest member of the series, methyllithium, is a non-volatile microcrystalline powder insoluble in hydrocarbons. Ethyllithium is a colourless crystalline compound melting at 95°. n-Propyl and n-butyllithium are almost colourless fairly viscous non-volatile oils soluble in hydrocarbons and ethers. These properties are to be compared with those of the corresponding sodium alkyls which are all colourless, non-volatile crystalline solids, insoluble in hydrocarbons. The difference in properties is usually attributed to differences in the type of bond between lithium and sodium alkyls, the former being considered covalent and the latter ionic compounds. Thus Coates (17) distinguishes between two types of compounds:

1) Essentially ionic compounds such as sodium alkyls, sodium phenyl etc. where nearly a full negative charge is localized on one carbon atom.

2) Essentially covalent compounds such as the lithium alkyls. Derivatives of *any* alkali metal where charge delocalization can occur over several carbon atoms (e. g. triphenylmethyl-sodium, lithium-benzyl) were also considered to be largely ionic in constitution.

It is possible that the lithium alkyls are more ionic in character than is commonly supposed and hence, that this distinction is somewhat artificial. Evidence based on solubility in non-polar solvents or the formation of a low melting solid structure are not necessarily convincing for alkyl ammonium halides often form crystals of low melting point and are soluble in benzene. Reference is often made to early estimates of Pauling which indicated only about 30% ionic character to the C–Li bond. Recent investigations with other compounds (21) have shown that these estimates lead to much too low values of ionic character. Methyllithium has all the characteristics of an ionic crystal (17) which would lead one to suppose that the higher members of the series cannot differ essentially in constitution. The low melting point of ethyl-lithium is due to the formation of a layer lattice (24) to accommodate the side chains, a phenomenon which is known for other salts such as ethylamine hydrochloride. The inability of the higher lithium alkyls to crystallize would seem to be mostly conditioned by geometric factors (t-butyllithium is a crystalline solid, n-butyllithium a liquid). There does however remain the problem that the sodium alkyls are all crystalline solids which suggests some special characteristics for the lithium compounds. It is a difficult problem to decide whether these are caused only by the small size of the lithium atom or if special bonding characteristics of lithium must be invoked.

In solution lithium alkyls are extensively associated especially in non-polar solvents. Ethyllithium in benzene solution exists largely as a hexamer (9, 43) in the concentration range down to 0.1 molar and there is no evidence for a trend with concentration so presumably the hexamers persist to even lower concentrations. Indeed even in the gas phase at high dilution it exists as hexamer and tetramer in almost equal amounts (3). In a similar way n-butyllithium in benzene or cyclohexane is predominantly hexameric (62, 122). t-Butyl-lithium however is mostly tetrameric in benzene or hexane (115). In ether solution both lithium phenyl and lithium benzyl exist as dimers (122) and it has been suggested that butyllithium behaves similarly in ether (15) although this does not agree with earlier cryoscopic measurements (122). It is however certain that more strongly basic ethers cause extensive breakdown of the structure.

b) Lithium alkenyls and aryls

These compounds are often strongly coloured in solution. In all cases the source is an absorption band centred in the visible or the near ultraviolet. These spectra are those expected for a carbanion involving extensive charge delocalization with minor perturbations due to the associated counter ion and to various extents of solvation. Simple HÜCKEL l. c. a. o. molecular orbital theory predicts such anions should have the same absorption band as the corresponding cation and at about the wavelengths observed although neglect of electron interaction makes exact correspondence unlikely. The growing centres in the polymerization of styrene, butadiene and isoprene are long chain metal alkenyls and aryls and show typical spectra. Only small changes in the position of maximum absorption are observed with different alkali metals and on change from a non-polar to a polar solvent. This suggests little change in the nature of the major species present other than differences of aggregation and solvation (48). As conductivity studies in tetrahydrofuran suggest ion-pairs as the major species (123) it is likely that these persist even in solvents of low dielectric constant although in extensively aggregated form (self solvation). The association of the polymerization intermediates is very similar to that observed for lithium alkyls. In non-polar solvents measurement of association is facilitated by the long chain nature of these compounds and in this case it is possible to show by viscosity and light-scattering methods that it persists down to very low concentrations.

III. Polymerization of non-polar monomers in non-polar solvents (benzene, hexane, heptane, cyclohexane)

a) Metallic lithium initiation

Metallic lithium in the form of a suspension has been used to polymerize isoprene (97) but the system is not too suitable for an exact analysis of the mechanism. The conversion-time curves are sigmoidal in shape. MINOUX (66) has shown that the overall rate is not very dependent on the amount of lithium dispersion used as expected if the organolithium intermediates are highly associated. The molecular weight of the polymer is more dependent on quantity of lithium used. The observed kinetic behaviour is very similar to that shown in lithium alkyl initiation. This suggests that apart from differences in the initiation step, the mechanisms are quite similar.

b) Lithium alkyl initiation (general)

Butyllithium is normally used as initiator although sometimes ethyllithium is used. There is so far no evidence that the different alkyls produce any significant changes in the initiation mechanism (25). With these initiators under rigorously pure conditions there is no evidence for termination or transfer reactions as the number average degree of polymerization is simply the ratio of initial monomer concentration to initiator concentration. Minor deviations (usually in the direction of too high measured molecular weights) can be attributed to catalyst loss by impurities. In addition viscosity-average or weight-average molecular weights are often determined which can be slightly higher than the number average if the chain initiation reaction is relatively slow. The active polymer chain ends show a characteristic ultraviolet absorption spectrum whose intensity remains stable at room temperature for extended periods giving another indication that the active species are quite stable to side reactions. The polymerization process consists therefore of simply a chain initiation and a chain propagation reaction and the molecular weight of the polymer should increase linearly with conversion. This has been found to be the case in the polymerization of isoprene (23).

The special characteristics of the polymerization process are connected with the association of organo-lithium compounds in solution. Early kinetic studies involved styrene as monomer. O'DRISCOLL and TOBOLSKY (81) suggested that the active initiator is the monomeric form of butyllithium in equilibrium with large amounts of inactive polymeric forms.

WELCH (*116*) also suggested association was prevalent and involved growing chains as well. In both cases the polymerization rate was observed to be nearly independent of initiator concentration above a critical value which was suggested to be the lower limit for association of the lithium alkyls (*116*). In both cases the detailed results are difficult to interpret as the monomer consumption curves are sigmoidal in shape, particularly so at high initiator concentrations. Measurements of the rate of polymerization determined as either the maximum rate (*81*) or from the apparently linear portion of a first order plot of monomer concentration (*116*) give an overall rate of polymerization which is a function of the propagation rate but also of the fraction of initiator molecules which have formed polymer chains. The initiation and propagation steps compete for monomer molecules and at high initiator concentrations it is easy to show that chain initiation will not be complete even when all the monomer has been consumed.

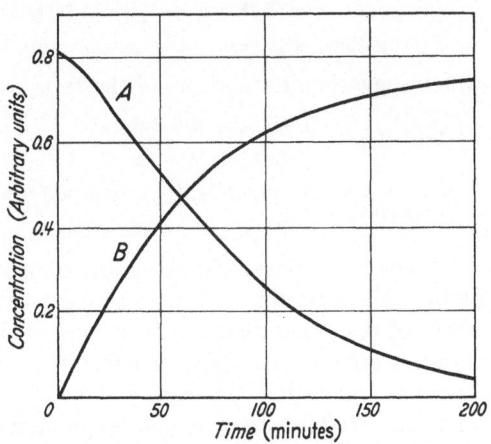

Fig. 1. Concentration of styrene monomer (A) and polystyryllithium (B) in arbitrary units during a polymerization of 1.4×10^{-2} molar styrene with 1.1×10^{-3} molar butyllithium in benzene

The styrene-benzene-butyllithium system was re-investigated by WORSFOLD and BYWATER (*124*) with a view to separate analysis of the initiation and propagation steps. This was achieved by simultaneous analysis for monomer concentration and concentration of active polymer chains. Both can be measured spectrophotometrically. Polystyryllithium in benzene has an absorption band at 334 mμ and styrene itself can be estimated from its absorption at 291 mμ. A typical experiment at high initiator concentration is shown in Fig. 1. Under these particular conditions the number of molecules of polystyryllithium is continuously increasing. At times of the order of 50 minutes the rate of increase in number of active chains is approximately compensated by the rate of decrease in monomer concentration and an almost constant rate of polymerization is observed. The rate of polymerization measured at this point has no real significance. Only after the addition of more monomer will chain initiation be complete as established from the constancy of the light absorption at 334 mμ. The rate of chain propagation can then be measured and correlated to a known number of active chains. At high monomer-initiator ratios the procedure is simpler as chain initiation is often complete before all the original styrene is consumed. In addition the initial rate of formation of polystyryllithium from curve B is equal to the rate of disappearance of butyllithium and gives the rate of chain initiation.

Analysis of this type can be carried out at various initiator and monomer concentrations and the kinetic orders can be established for both reaction steps. For the case under consideration these are:

Initiation: $-d\,[\text{BuLi}]/dt = k_1\,[\text{BuLi}]^{0.15}\,[\text{styrene}]^{1.0}$,

Propagation: $-d\,[\text{styrene}]/dt = k_2\,[\text{polystyryllithium}]_{total}^{0.5}\,[\text{styrene}]^{1.0}$,

which suggests a mechanism of the type:

$$
\begin{array}{ll}
\text{Initiation} & \left\{ \begin{array}{ll} (\text{BuLi})_6 \leftrightharpoons 6\,\text{BuLi} & K_1 \\ \text{BuLi} + \text{M} \rightarrow \text{Bu}(\text{M}_1)^-\text{Li}^+ & k_i \end{array} \right. \\[2ex]
\text{Propagation} & \left\{ \begin{array}{ll} (\text{Bu}(\text{M})_n^-\text{Li}^+)_2 \leftrightharpoons 2\,\text{Bu}(\text{M})_n^-\text{Li}^+ & K_2 \\ \text{Bu}(\text{M})_n^-\text{Li}^+ + \text{M} \rightarrow \text{Bu}(\text{M})_{n+1}^-\text{Li}^+ . & k_p \end{array} \right.
\end{array}
$$

This scheme requires the assumption of extremely strong association of all lithium-organics down to at least 10^{-4} molar concentration if the observed reaction orders are to be obeyed. It assumes in agreement with earlier workers that only unassociated lithium alkyls and aryls are reactive. The six-fold association of butyllithium required is in agreement with physical measurements although admittedly these were carried out at much higher concentrations. Morton and co-workers (69) have shown that the polymer molecules are indeed associated into dimers in this system from a quantitative study of the decrease in solution viscosity on removal of the charged species at the ends of the polymer molecules.

The above analysis method measures the propagation rate only when all butyllithium has reacted and the initiation rate when very few polymer chains are present. It is possible that in the intermediate stages mixed associated species are present involving butyllithium and polymer chains. In the polymerization of styrene a value of k_2/k_1 can be derived from an approximate treatment of the overall process which is in reasonable agreement with the separately determined values of k_1 and k_2 so that mixed species do not have an important effect on the overall kinetics. This may not be true however in other cases.

The observed rate constants k_1, k_2 are obviously complex quantities each of which involves a rate constant and an equilibrium constant, $k_1 = k_i K_1^{1/6}$, $k_2 = k_p K_2^{1/2}$. It is not possible to separate the individual contributions from the present measurements.

c) The propagation reaction

The butyllithium-styrene-benzene system is the only one to the present time in which it has been possible to carry out such a detailed analysis of the reaction mechanism. Usually only the propagation step has been studied. Two methods (in addition to the one described in the previous section) have been used to ensure that chain initiation is complete and hence that the added initiator concentration is equal to the number of reactive polymer chains. 1) Analysis of a sample from the polymerization mixture for butane after deliberate termination with a suitable proton donor ($BuLi + H^+ \rightarrow BuH + Li^+$). This method enables

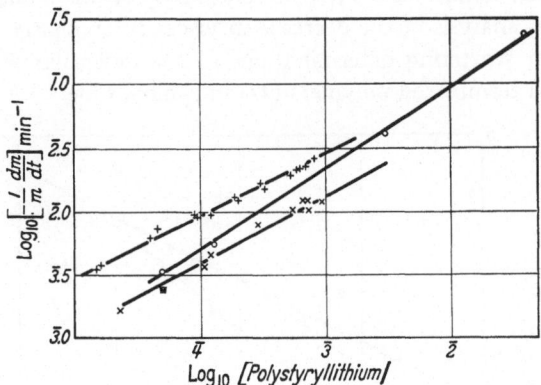

Fig. 2. Variation of the propagation rate with total concentration of polystyryllithium. Results of: + WORS-FOLD and BYWATER in benzene, O SPIRIN et al. in toluene, × JOHNSON and WORSFOLD in cyclohexane

the point to be found at which no free butyllithium exists. 2) Equilibration of the initiator with a small amount of monomer before carrying out a polymerization ("seeding technique"). It is then assumed that all the butyllithium has reacted. This is not always necessarily true as, owing to the peculiar characteristics of these polymerizations, often rather large amounts of monomer are required.

The results for styrene polymerization in non-polar solvents at 30° are summarized in Fig. 2 which is derived from the results of WORSFOLD and BYWATER (124), SPIRIN, GANTMAKHER and MEDVEDEV (100, 101) and JOHNSON and WORSFOLD (47). MORTON, FETTERS and BOSTICK'S results in benzene presented graphically in reference (73) seem to be in good agreement with the results in the figure. The ordinate represents the first order rate of monomer consumption $\left(-\dfrac{1}{[M]}\dfrac{d[M]}{dt}\right)$ plotted logarithmically. The slope of the line therefore defines the exponent (x) in the relationship $-\dfrac{1}{[M]}\dfrac{d[M]}{dt} = k_2$ [total polystyryllithium]^x. This is seen to be close to 0.5 in benzene and cyclohexane but a little higher in toluene (0.6). The latter deviation could be due to some initiator destruction at low concentrations, for one would expect the results in toluene and benzene to be closely similar. In cyclohexane the experiments were actually carried out at 40° as the polymer is insoluble at 30°. 0.4 log. units have been subtracted from the logarithm of the true rate at 40° to give an approximate correction to 30°. The overall propagation rate is however undoubtedly lower in cyclohexane but it is not clear if this is due to change in K_2, k_p or both. All the results are consistent with the idea that most of the polystyryllithium is associated to dimers which are unreactive in the polymerization. Dimerization by interaction of the ion-pairs has been confirmed in benzene solution by MORTON and co-workers (69) from an analysis of the decrease in viscosity on charge destruction, and by light scattering determinations of the molecular weight of the "living" and terminated polymers (47) in cyclohexane.

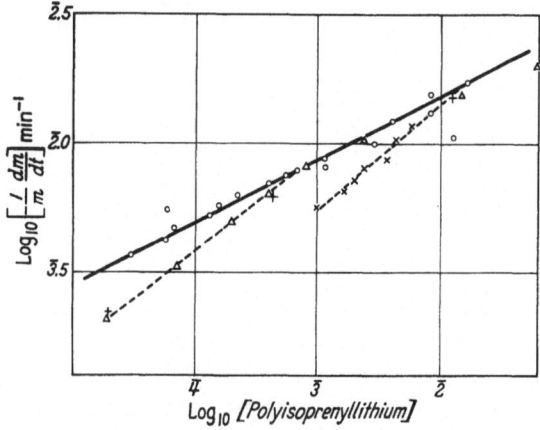

Fig. 3. Variation of the propagation rate with total concentration of polyisoprenyllithium. Results of: O WORSFOLD and BYWATER in cyclohexane, × MORTON et al. in hexane, + SPIRIN et al. in heptane, △ SINN et al. in heptane

Fig. 3 shows a comparison of results for isoprene from the work of SINN (96), MORTON (72), SPIRIN (100) and WORSFOLD and BYWATER (126). They have been corrected to 30° wherever necessary using the authors reported activation energies. The results are in good agreement at high polyisoprenyllithium concentrations but deviate as the concentration is decreased. The experiments of SINN (heptane) and WORSFOLD and BYWATER (cyclohexane) are in excellent agreement down to a concentration of 3×10^{-4} molar. Below this concentration the former authors' results show a progressive drop in rates which is most marked below 10^{-5} molar concentrations (section III d). The results of SPIRIN (heptane) follow those of SINN fairly closely throughout the range. MORTON's results (hexane) diverge progressively from the rest at lower initiator concentrations.

The majority of the results are therefore described by the expression
$$-\frac{1}{[M]}\frac{d[M]}{dt} = k_2 \text{ [polyisoprenyllithium]}_{total}^{0.25}$$ which suggests that the

polymer molecules in this concentration range exist mainly as tetramers. An association number of about four was confirmed by light-scattering measurements in cyclohexane (126) and in heptane at millimolar concentrations by a viscosity method (94). MORTON's results suggest dimeric association, an interpretation which he confirmed by experiments on the relative solution viscosities of active and terminated polymers. There appears to be a genuine difference in degree of association between MORTON's results and the others. It does not seem likely that this should be caused by the use of hexane as solvent rather than heptane or cyclohexane. The most plausible explanation for the differences observed with isoprene is that the rates and degree of association are very sensitive to the presence of small amounts of impurities produced by initiator destruction.

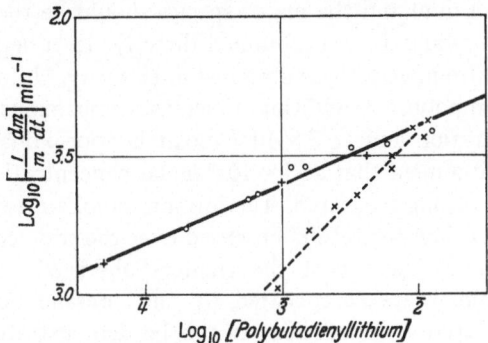

Fig. 4. Variation of the propagation rate with total concentration of polybutadienyllithium. Results of: ○ JOHNSON and WORSFOLD in cyclohexane, + SPIRIN et al. in heptane, × MORTON et al. in hexane.

The propagation reaction has been studied also for butadiene, the results being shown in Fig. 4. Once again the results of SPIRIN (100) (heptane) and WORSFOLD (47) (cyclohexane) are in good agreement whereas the results of MORTON (69, 72) diverge at lower initiator concentrations. The former results are described by the expression $-\dfrac{1}{[M]}\dfrac{d[M]}{dt}$ $= k_2$ [polybutadienyllithium]$_{total}^{1/6}$. The kinetic order in MORTON's experiments is between one-half and one-third. His measurements of association number by the viscosity method confirm that this corresponds to a degree of association between two and three. An explanation for the observed differences can be made in exactly the same manner as for isoprene. If we assume the 1/6th order is correct this indicates that polybutadienyllithium molecules are more heavily associated than those of polyisoprenyllithium. This is in fact a similar difference to that found between n-butyl and t-butyllithium and must be ascribed to steric effects.

d) Determination of absolute rate constants for chain propagation

The observed rate constants for chain propagation (k_2) are functions of k_p and the equilibrium constant K_2. It would be of great interest to obtain k_p itself. This can be done in principle if conditions are chosen where association is negligible or alternatively if the association constant can be measured independently. All data obtained so far apply to isoprene polymerization.

Sinn (95, 96) has investigated the propagation rate at polyisoprenyl-lithium concentrations as low as 5×10^{-7} molar. The reaction order in polyisoprenyllithium was found to increase at concentrations below 10^{-4} molar. (The beginning of this effect is visible in Fig. 3.) At a concentration of 5×10^{-6} molar the rate was about first order in polyisoprenyl-lithium and the major species should be the free ion-pair. The absolute k_p value is then obtained directly. In n-heptane at 20° a value of 0.65 litre/mole sec was obtained in this way. Unfortunately the reaction order in polyisoprenyllithium increases continuously below 10^{-4} molar concentration and at 5×10^{-7} molar is approximately two. This weakens the argument that at 5×10^{-6} molar concentration a simple ion-pair reaction is being measured. The authors invoke a rather complex mechanism to explain the results observed over the wide concentration range. It seems more likely that the changes observed at concentrations below 10^{-4} molar are complicated by the presence of lithium salts formed by initiator destruction. It can be demonstrated that in a greaseless high vacuum system, under conditions where the glassware as well as solvent and monomer are pre-treated with butyllithium or its equivalent, the concentration of reactive impurities is normally about 3×10^{-6} molar. Less rigorous procedures will naturally lead to more initiator loss. Experiments carried out at initiator concentrations below 3×10^{-5} molar cannot therefore be considered to be accurate even under the most favourable conditions. A correction for the loss of lithium alkyl can be made, but at 3×10^{-6} molar polyisoprenyllithium the reaction is being carried out in the presence of at least an equal quantity of lithium salts which will have an appreciable effect on measured rates. Roovers and Bywater (89), for instance, have found that t-butoxylithium depresses the propagation rate of polystyryllithium in benzene. It is clearly desirable to obtain more general information on the effect of lithium salts on this type of reaction in view of the above considerations.

An alternative method has been used by Morton (73, 74) which involves an independent measurement of the degree of association of polyisoprenyllithium. Measurements were made of the concentrated solution viscosity of active and terminated polymer solutions. An approximately ten-fold decrease in viscosity was observed on discharging the ion-pair. This corresponds to two-fold association for polyisoprenyl-lithium in hexane, since in concentrated solutions $\eta = K \overline{M}_W^{3.4}$. A careful study established that the association numbers were slightly less than two and decreased with temperature. From the data, K_2 can be calculated and hence k_p. In this way it was found that in hexane $k_p = 3.4 \times 10^3$ exp. $(-4100/RT)$; at 30°, $k_p = 4.7$ litre/mole sec. For practical reasons, the accuracy is somewhat limited. At 40° for instance an error in the viscosity exponent of only 0.02 will cause an error in K_2 of about 70%. The method is also extremely sensitive to errors in the determination of the association number N. K_2 is determined as $2(\text{RLi})_0(2 - N)^2/(N-1)$ and N is close to two (1.964 at 40°). A more serious difficulty arises because other workers find a much higher degree of association and this must give rise to some doubt as to the accuracy of the k_p value derived.

A third attempt at absolute k_p measurement has been made for isoprene in cyclohexane (30). The value obtained was 0.03 litre/mole sec at 15°. The experiments involved polymerization in the presence of ethyllithium. The kinetic analysis used implicitly assumes that under all conditions only free polyisoprenyllithium and its dimer with ethyllithium exist in appreciable concentrations. With this assumption, k_p and K_2 can be determined from the variation in rate with the relative concentrations of the two species. The equilibria involved are undoubtedly more complex than this and experiments of this type are unlikely to give a true value for k_p.

The problem is an extremely difficult one to solve owing to the high interaction energy of lithium compounds in non-polar solvents and their extreme reactivity to impurities. In the experiments of WORSFOLD and BYWATER, the reaction order is maintained down to a polyisoprenyl-lithium concentration of 3×10^{-5} molar. If it is assumed that this requires that less than 10% of unassociated material be present, then k_p must be greater than about 20 litre/mole sec at 30°. All the measured k_p's (Table 1) are lower than this and the author feels that despite the ingenuity shown by various authors no accurate value for isoprene has yet been obtained.

e) The initiation reaction

This has been studied much less frequently and appears to be a rather more complex reaction. The first results obtained, for the butyl-lithium, styrene reaction in benzene have already been described. In a similar way the addition of butyllithium to 1,1-diphenylethylene shows identical kinetic behaviour in benzene (26). Even the proton extraction reaction with fluorene shows the typical one-sixth order in butyllithium (27). It appears therefore that in benzene solution at least, lithium alkyls react via a small equilibrium concentration of unassociated alkyl. This will of course not be true for reactions with polar molecules for reasons which will be apparent later. No definite information can be obtained on the dissociation process. It is possible that the hexamer dissociates completely on removal of one molecule or that a whole series of penta-mers, tetramers etc. exist in equilibrium. As long as equilibrium is maintained, the hexamer is the major species present and only monomeric butyllithium is reactive, the reaction order will be one-sixth. A plausible suggestion can be made in which the dissociation proceeds via small concentrations of tetramer and dimer. There is some evidence that these forms have some degree of stability (3, 24).

It would be natural to assume that in aliphatic solvents similar processes would occur. Attempts to study the polymerization initiation reaction in cyclohexane however have not confirmed that this is so (47, 126). In this solvent with either styrene, isoprene or butadiene as monomer a more complex mechanism is indicated. In all cases, chain initiation, as measured by the spectrophotometric determination of polymer anion concentration appears to start slowly and to gradually increase in rate. Instead of the instantaneous and initially linear increase in active centre concentration observed on mixing the reactants in benzene (Fig. 1) the concentration-time curve is sigmoidal in shape. It resembles the typical monomer consumption curve observed in these systems. The exact cause of this behaviour is not clear. The effect could be caused by residual impurities in the system and in fact this possibility cannot be entirely eliminated. However, exactly the same stringent procedures of purification of materials and treatment of glassware were used as with benzene. In addition, at least with isoprene, the deliberate addition of trace amounts of air accelerates the initiation rate, completely eliminating the period of slow rate increase. Less stringent drying of butadiene produces the same effect. These results suggest that the variable initiation rate is not caused by impurities but is a characteristic of the most rigorously purified systems.

The difference in behaviour between benzene and cyclohexane is surprising. It is generally accepted however that benzene is a much better solvating agent than is an aliphatic solvent and this factor could cause the equilibrium concentration of free butyllithium in benzene to be much larger than in cyclohexane. The direct attack of free butyllithium on monomer would then proceed at a sufficient rate in benzene that competing processes could be negligible. In cyclohexane, this process appears to be very slow and in the presence of small amounts of polymer ion-pairs or deliberately added lithium salts a faster alternative process determines the initiation rate. Further work on the initiation reaction is obviously necessary before its general nature can be understood.

f) Effect of traces of polar materials
on polymerization in hydrocarbon solvents

Early workers in the field (53, 81, 116) discovered that polar materials (e. g. ethers, amines), present in amounts comparable to the initiator concentration led to a large increase in the polymerization rate of styrene in non-polar solvents. This has since been found to be quite a general phenomenon. Since only overall polymerization rates were measured, however, it was not clear whether it was due to acceleration of chain initiation or to chain propagation or both. KROPACHEVA and colleagues (59) showed that tetrahydrofuran markedly increased the initiation rate. The major effect in fact seems to be to cause a rapid increase in initiation rate (14) and, except at very low concentrations, chain initiation becomes too rapid in the presence of tetrahydrofuran to measure by conventional means.

Few detailed studies have been made on the effect of Lewis bases on the propagation rate, but the effect of tetrahydrofuran is known for

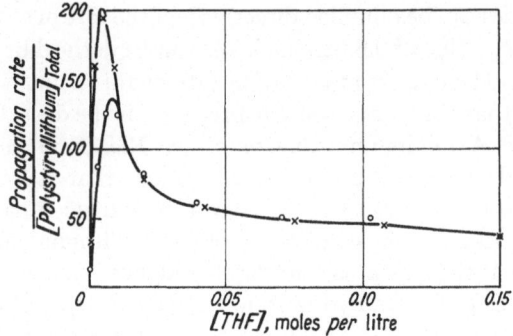

Fig. 5. Effect of tetrahydrofuran on the propagation rate in the polymerization of styrene in benzene. Concentration of polystyryllithium, ○ ~1.1 × 10⁻³ molar, × ~1.4 × 10⁻⁴ molar

isoprene and styrene. It is much smaller than observed for chain initiation. With styrene in benzene (14), the propagation rate shows rather complex behaviour as tetrahydrofuran is added, as shown in Fig. 5. The rate first increases with increasing THF concentration and subsequently decreases to a plateau where the rate per polystyryllithium molecule is independent of its concentration. On the ascending part of the curves, the reaction is still one-half order in polystyryllithium as in pure benzene. Above THF concentrations about .02 molar the propagation rate is clearly first order. This must be interpreted as a complete breakdown of the dimeric polymer ion-pairs at a relatively low THF concentration, and their replacement by tetrahydrofuran solvated ion-pairs. A more detailed kinetic analysis (14) reveals that a plausible explanation for the results will involve the presence of at least four different polymer species in labile equilibrium: the normal polymer ion-pair dimer $(RM_xLi)_2$, its dissociation product RM_xLi, a mono-etherate $RM_xLi \cdot THF$, and a dietherate $RM_xLi \cdot 2THF$. The ion-pair dimer is as usual assumed to be inactive but the other species must be active in polymerization but in varying degree. The initial increase in rate is caused mainly by the presence of increasing amounts of $RM_xLi \cdot THF$

which is relatively active. At this point, the reaction order is still one-half in total concentration of polystyryllithium. $(RM_xLi)_2$ is still therefore the major species present and the concentration of RM_xLi cannot change much if the equilibria between the species are rapidly established. At higher THF concentrations the equilibria are gradually displaced over to the dietherate, which must be less active in polymerization, and the rate decreases. The k_p values for RM_xLi and $RM_xLi \cdot THF$ cannot be determined but only a complex constant which as usual involves the equilibrium constant. It can be assumed that at about 0.15 molar tetrahydrofuran all the polystyryllithium is in the form $RM_xLi \cdot 2THF$ and k_p for this species can be obtained directly from the rate of chain propagation. At 20° a k_p value of 0.6 litre/mole sec can be derived in this way for the dietherate. The overall propagation rate at this point is not greatly different from that observed in pure benzene nor is the dielectric constant of the solvent. A minimum value of k_p for RM_xLi in benzene is 30 litre/mole sec. This value is obtained by assuming that at 2×10^{-5} molar total concentration there must be less than 10% of free ion-pair present to maintain the half order dependence on polystyryllithium concentration. It is possible that the true k_p is considerably higher and equals or exceeds that observed in pure tetrahydrofuran. Strong solvation of the ion-pair obviously decreases its reactivity to a large degree. Even in benzene the ion-pair must be solvated to some extent but the association will be much weaker.

The effect of tetrahydrofuran on the polymerization of isoprene in hexane has been studied by Morton and co-workers (73, 74). The viscosity method was used to measure the degree of association. This was found to drop from 2 in pure hexane to about 1.3 with a ratio of THF to polyisoprenyllithium of 100 and dissociation of the polymer aggregates was complete at ratios of 500—700. With the reasonable assumption that the only species present in significant amounts were associated polymer molecules and etherates, it was possible to find the concentration of etherate present under all conditions. An equilibrium constant could be evaluated from the overall process

$$(RM_xLi)_2 + 2n \cdot THF \leftrightharpoons 2RM_xLi \cdot nTHF .$$

The results were consistent only with a value $n = 1$. Hence apparently in this case a monoetherate only is formed. The heat of association of tetrahydrofuran with the ion-pair is found to be 18 kilocal/mole, a value about that expected from simple electrostatic calculations. At low THF concentrations where the dielectric constant is not appreciably different from that in pure hexane, k_p for the etherate was evaluated as 0.14 litre/mole sec at 30°. The overall propagation rate increases steadily up to a tetrahydrofuran concentration of about 1 molar and then decreases steadily with increasing tetrahydrofuran concentration. This decrease is unexpected and its exact cause is not known at the present time.

Similar results were obtained by WORSFOLD and BYWATER (126) in cyclohexane. At about millimolar concentrations of polyisoprenyllithium the propagation rate was found to double on the addition of tenth molar tetrahydrofuran. The reaction order in polyisoprenyllithium increased from one quarter to one half in the same range. This indicates that 0.1 molar tetrahydrofuran is not sufficient to dissociate the polymer ion-pair aggregates. With styrene, dissociation is complete at 0.02 molar tetrahydrofuran. No maximum is observed in the propagation rate as the amount of tetrahydrofuran is increased. The results of both observers are thus in qualitative agreement that evidence exists only for a single etherate which is much less readily formed than with polystyryllithium.

IV. Polymerization of polar monomers in non-polar solvents

a) Methylmethacrylate in toluene

The mechanism of the polymerization of this monomer has been studied in far greater detail than any other. It is clear from the outset that a much more complex mechanism is involved than is the case for olefins. A large proportion of the initiator is used to form polymer whose molecular weight is only a few hundreds and the overall molecular weight distribution is so broad as to be rivalled only by those found in polyethylene produced by the high pressure process (19, 39). The initiator disappears almost instantaneously on mixing the reactants (19, 38). Under these conditions, an almost monodisperse polymer would be expected if chain transfer or termination processes are absent.

GLUSKER (37, 38) attempted to prove that these processes are absent by an estimation of active chains by reaction with C^{14} labelled CO_2 or $H^3(T)$ labelled acetic acid, followed by measurements of the radioactivity of the polymer isolated. Most of the experiments were carried out with fluorenyllithium as initiator in toluene containing 10% diethylether at $-60°$. At $-78°$ at least 80% of the polymer chains were found to be active at the end of polymerization. The lowest fraction was appreciably less active. Similar results were obtained at $-60°$ although no examination was made of the fractions of lowest molecular weight. Kinetic experiments indicated a first order decay of monomer concentration after an initial rapid consumption of about 3 molecules of monomer per initiator molecule. The mechanism suggested to explain these results involves rapid addition of fluorenyllithium across the vinyl double bond followed by the rapid addition of three monomer units. At this stage it is suggested most of the growing chains are immobilized as I where R is fluorenyl.

Structures such as I, which are to be regarded as temporarily immobilized by interaction of the cation with the carbonyl group two monomer units away, can also occur for longer polymer chains, i. e., where R represents a polymer chain. Reaction of the cyclic complex with $A\bar{c}OT*$ was supposed to proceed in two different ways to form II and III. Only product II has the tritium atom on the polymer chain, and hence low polymer would have a lower tritium count as actually observed. If a constant fraction of growing chains were immobilized after the addition of each monomer molecule, reaction would virtually cease after a very few monomer units had added. It is necessary to postulate that after a chain length of about ten units is attained there is a considerably decreased probability of pseudo-cyclization. The authors originally considered that at this point a helical conformation can be formed which favours isotactic placements and has a low cyclization probability. Later evidence in this and other systems (19, 36) has shown that in toluene the very low molecular weight polymer is isotactic and that under conditions where the polymer formed is highly syndiotactic large amounts of hexane-soluble polymer are still formed. This hypothesis must therefore be abandoned. The formation of large amounts of hexane-soluble polymer can only be correlated with abnormal conditions existing in the first few seconds of reaction. The exact cause is unknown. The first order rate of monomer disappearance established after a few seconds requires that a steady concentration of active and non-active chains is established. The cyclic complex must therefore have a small probability of reacting with monomer and growing further.

Butyllithium initiation of methylmethacrylate has been studied by KOROTKOV (55) and by WILES and BYWATER (118). Korotkov's scheme involves four reactions: 1) attack of butyllithium on the vinyl double bond to produce an active centre, 2) attack of butyllithium at the ester group of the monomer to give inactive products, 3) chain propagation, and 4) chain termination by attack of the polymer anion on the monomer ester function. On the basis of this reaction scheme an expression could be derived for the rate of monomer consumption which is unfortunately too complex for use directly and requires drastic simplification. The final expression derived is therefore only valid for low conversions and slow termination, and if propagation is rapid compared to initiation. The mechanism does not explain the initial rapid uptake of monomer observed, nor the period of anomalous propagation often observed with this initiator. The assumption that $k_p \gg k_i$ is hardly likely to be true even after allowance is made for the fact that the concentration of active species is much smaller than that of the added initiator. Butyllithium disappears almost instantaneously but propagation proceeds over periods from tens to hundreds of minutes. The rate constants finally derived therefore cannot be taken seriously (the estimated k_p is 2×10^5 that of k_i) nor can the mechanism be regarded as confirmed.

WILES and BYWATER (118) found that the polymerization process with butyllithium was much too complex to fit a simple scheme. The polymerizations were carried out in toluene at $-30°$. An appreciable fraction of the monomer was always consumed extremely rapidly. The product at this stage was mostly low molecular weight polymer, soluble in petroleum ether, and of molecular weight about 800. Some polymer of higher molecular weight was also present even at this stage. At low initiator concentrations, the initial rapid consumption of monomer was followed by a slow period of growth but after about 25% of the monomer had been consumed the polymerization rate increased and a roughly first order disappearance of monomer followed. At high initiator concentration the slow phase was not noticeable but could have been masked since the overall reaction was quite rapid. Addition of further monomer after the complete polymerization of the first batch caused a first order monomer consumption closely equal to that established in the later stages of polymerization of the first monomer sample. No additional low molecular weight material was produced. It appears therefore that a steady concentration of active chains is produced in later reaction stages. The high molecular weight polymer accounted for less than 10% of the initiator molecules. A larger fraction existed as petroleum ether-soluble polymer but all the butyllithium added could not be accounted for if it was assumed that each polymer molecule contained one butyllithium residue (19). In some cases the apparent loss was as high as 60%. This type of result means that either i) multiple attack of butyllithium has occurred, ii) some initiator was lost due to reaction with the ester function of the monomer, or iii) very low polymer (e. g. dimer) was lost by evaporation in the isolation procedure.

By fractionation of products formed at various conversions and on multiple addition of monomer other facts could be established. Thus the bimodal distribution is formed early in the reaction and persists throughout. Growing polymer chains of all chain lengths exist with the possible exception of the petroleum ether-soluble material, which is essentially inactive at least by the end of polymerization. The longer chains seem to be adding monomer faster.

A possible source of at least part of the complex behaviour observed would be the presence of lithium methoxide in polymerization mixtures. This could be formed by reaction of butyllithium with methylmethacrylate,

$$
\begin{array}{ccc}
CH_3 \quad O & CH_3 \quad C_4H_9 & CH_3 \quad C_4H_9 \\
\diagdown C-C \diagup + C_4H_9Li \rightarrow \diagdown C-C-OLi \rightarrow \diagdown C-C \diagup + LiOCH_3, \\
\diagup CH_2 \quad \diagdown OCH_3 & \diagup CH_2 \quad \diagdown OCH_3 & \diagup CH_2 \quad \diagdown O
\end{array}
$$

or via the cyclization of polymer chains indicated earlier, or by interaction of a growing polymer chain with monomer

$$
\begin{array}{ccc}
O \quad CH_3 & OLi \quad CH_3 & O \quad CH_3 \\
P_x^- + \diagdown C-C \diagup \rightarrow P_x-C-C \diagup \rightarrow P_x-C-C \diagup + LiOCH_3. \\
\diagup CH_3O \quad \diagdown CH_2 & OCH_3 \quad \diagdown CH_2 & \diagdown CH_2
\end{array}
$$

Evidence for such reactions in methylmethacrylate polymerizations was obtained by termination of polymerizations with acetic acid followed by measurements of methanol formed (119, 120). Fig. 6 shows typical results obtained. It is assumed that the methanol found corresponds to lithium methoxide in the reaction mixture. Some methanol might in fact be produced only in the termination reaction from pseudo-cyclized species (cf. p. 82). In addition the tertiary alkoxides formed by attack of the initiator or growing polymer chains on the carbonyl group of the monomer might not eliminate lithium methoxide immediately but would do so on termination with acetic acid. In any case much of the methanol formed corresponds to actual alkoxide in the reaction mixture and the results give a minimum value of the concentration of species inactive in polymerization. For brevity it will be referred to as lithium methoxide.

With butyllithium most of the methoxide is produced in the first few seconds at $-30°$. The rapid formation must then correspond to attack of butyllithium on ester groups or to the termination (or pseudo-termination) step involved in the formation of low polymer. The amount formed is greater than the number of molecules of low polymer so not all of the methoxide is formed from cyclization. If diphenylhexyllithium is used as initiator at $-30°$ virtually no methoxide is formed in the first few seconds where appreciable low molecular weight polymer is produced (Fig. 6). This suggests that most of the initial methoxide with butyl-

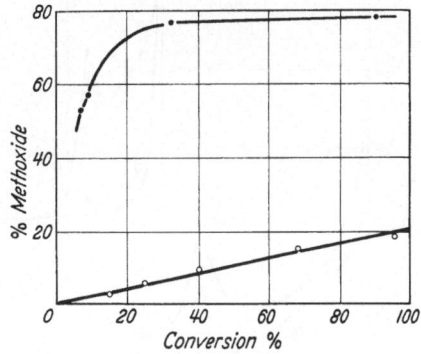

lithium comes from initiator attack on the ester group of the monomer. If this is so, the low molecular weight polymer cannot be terminated by a true cyclization. If it is in the form of the cyclic complex then it must predominantly react with acetic acid at $-30°$ so as not to cyclize. At $-78°$ and in the presence of 10% ether it was necessary to assume that an appreciable amount of the low molecular weight product cyclized on reaction with acetic acid. With diphenylhexyllithium

Fig. 6. Formation of lithium methoxide in the polymerization of methylmethacrylate (0.125 molar) in toluene at $-30°$. Initiator: ○ 3.2×10^{-3} molar diphenylhexyllithium, ● 3.85×10^{-3} molar butyllithium

at $-30°$, the methoxide concentration is lower and increases linearly with conversion. With butyllithium there is also a slower phase of methoxide elimination following the first rapid stage. Although some of this could be caused by termination of long chains much appears to be formed from the petroleum ether-soluble polymer which must slowly undergo a real termination reaction at $-30°$. This reaction is quite temperature dependent and would be less noticeable at $-60°$.

It is likely that some and perhaps all the methanol measured corresponds to the presence of lithium methoxide. This must have some influence on the reaction. Experiments show that it in fact accelerates the polymerization when deliberately added even though it does not initiate polymerization under the conditions used. Even more effective are lithium ethoxide and propoxide which are considerably more soluble in toluene. Lithium methoxide itself is virtually insoluble in toluene, but produced in situ is probably solubilized by association with the active polymer chain ends. Some of it, however, might not be in solution and the possibility of some reaction occurring on the surface of colloidally dispersed salt cannot be excluded. It is interesting to note that both initiators (fluorenyllithium and butyllithium) which produce a high terminal fraction of "lithium methoxide" give bimodal distributions, whereas diphenylhexyllithium does not (Fig. 7).

Wiles and Bywater (*120*) considered that all the measured methanol probably corresponds to actual lithium methoxide present in the reaction. It was considered that the bimodal distribution is caused by its presence, at least two different types of active chain being present, those uncomplexed and those complexed with lithium methoxide. It is necessary to assume that rapid exchange between the two types does not occur and that each adds monomer at its own characteristic rate. Some chain termination at least in early reaction stages would also serve to broaden the distribution.

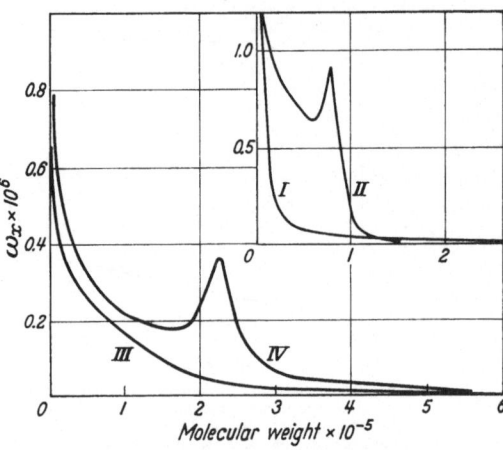

In summary, it is clear that polymerization of methylmethacrylate follows a complex pattern. In all cases, a large fraction of the polymer chains is rapidly immobilized at very short chain lengths. The formation of the cyclic complex could account for this behaviour, for these chains

Fig. 7. Typical differential weight distribution curves observed in the polymerization of methylmethacrylate in toluene. Initiated by: I Phenylmagnesiumbromide, II Fluorenyllithium, III Diphenylhexyllithium, IV Butyllithium

do not seem to be really terminated at least at the beginning of the reaction. Some true termination can occur as reaction proceeds but its extent depends on temperature and possibly on the presence of added ethers. There are however obvious difficulties in assignment of the cyclic complex structure to pseudo-terminated chains. No adequate reason exists why it should only form readily at short chain lengths. The plausibility of its existence depends on the identification of a cyclic ketone in low molecular weight product after reaction with acetic acid. This identification depends largely on assignments of infra-red absorption bands whereas there is lack of agreement as to the spectra of such ketones (*86, 114*). We cannot exclude the possibility therefore that another unknown mechanism is responsible for the phenomenon of pseudo-terminated chains.

The broad distribution of molecular weights can be explained by equilibration of active and pseudo-terminated chains as Glusker has suggested. It is not clear if a general mechanism of this type would predict the exact form which is observed experimentally or if it could account for specific differences observed for instance between butyl-lithium and diphenylhexyllithium initiated products. The abnormal behaviour observed with butyllithium is probably connected with attack on the ester group of the monomer to produce immediately large amounts of alkoxides. The other initiators still produce some alkoxide and this can affect the active centres and influence the molecular weight distribution, although perhaps not as markedly if ethers are present. In addition, a small amount of slow termination of the active chains still cannot be completely excluded and this again would serve to broaden the distribution.

Kinetic studies on the fluorenyllithium-initiated polymerization reveal a rather complex behaviour under many conditions (36). Experiments in toluene were always carried out in the presence of some amount of ether or tetrahydrofuran. The simplest behaviour occurs with the systems containing 7–20 volume-% of ether. Here the polymeriza-tion rates are independent of ether concentration and are first order in monomer and fluorenyllithium. As most of the initiator is consumed in the rapid formation of low molecular weight polymer, this implies that there is a simple relationship between initiator concentration and active chain concentration [XLi]. Evidence was provided that in all experi-ments containing ether, 17.6% of the initiator produced active chains whatever the ether or monomer concentration. On this basis, the propaga-tion reaction is postulated to occur by direct reaction with monomer of the chains which survive the rapid changes producing low molecular weight polymer. As the ether concentration is lowered below 7% the reaction tends towards second order in initiator and it is assumed that dimerization of active species becomes appreciable and that at 2% ether reaction proceeds entirely through the dimerized species. This will produce a reaction which is second order in added initiator provided that XLi is still the major species present, equilibrium is rapidly establish-ed, and the intrinsic reactivity of the dimer is much greater than that of XLi.

Experiments in the presence of tetrahydrofuran show that the polymerization rates are always proportional to the square root of added initiator concentration (0.1% to 15% THF). 34% of the initiator is found to lead to active chains. The dependence on monomer concentra-tion is strange. The monomer reacts according to a first order law but the rates are usually dependent on initial monomer concentration. The square root dependence on initiator concentration can be explained if reaction proceeds via free (solvated) ions in labile equilibrium with the undissociated solvated species XLi · nTHF, if the latter is still the major species present. The mechanisms suggested formally explain the observed kinetic orders but cannot be considered as being entirely satisfactory. It seems improbable that dimerized species should be more reactive than the unassociated species. There is evidence that free anions have a greatly enhanced reactivity but it is doubtful that at low concentrations of tetrahydrofuran, where the dielectric constant of the solvent is virtually that of toluene, that there would be sufficient free ions present to contribute appreciably to the reaction rate. The results strongly suggest that the relationship between active chain concentration and the initiator concentration is not always as simple as the authors propose. In some cases, this could provide an explanation of the observed orders without the necessity for postulation of reactions other than directly with the active species. At low tetrahydrofuran concentrations for instance, the results are equally compatible with an active chain concentration which increases less rapidly than the increase in added initiator, and is a function of the tetrahydrofuran concentration. The inevitable scatter of experimental points in this type of work makes a definite conclusion impossible, but this type of explanation should not be discarded lightly. The fact that the authors report different initiator efficiencies in the presence of ether and tetrahydrofuran would suggest that as the concentration of additive reaches low values the efficiency would probably change to a value characteristic of pure toluene.

With diphenylhexyllithium (121) (the product of addition of butyllithium to 1,1-diphenylethylene) kinetic results are the same as found for fluorenyllithium initiation in the presence of moderate amounts of ether. Even in pure toluene, the rates are first order with respect to initiator concentration and monomer concentration. This simple behaviour is caused by a constant fraction of the initiator forming low molecular weight polymer. If butyllithium is used as initiator, the kinetic behaviour is too complex for analysis.

b) Other polar monomers in non-polar solvents

A number of other polar monomers have been polymerized with butyllithium, nominally in hydrocarbon or aromatic solvents. In almost all cases the monomer concentration was so high that the effective dielectric constant was much greater than in a pure hydrocarbon. All show rather complex behaviour. The degree of polymerization of the polymer formed is always much higher than the initial monomer-catalyst ratio so that a simple scheme involving only initiation and propagation reactions is not applicable. Only precipitable polymer was isolated, so it is not sure if the low initiator efficiencies are due to low polymer formation or to side reactions of butyllithium with the monomer. In addition most systems studied stop before complete conversion of the monomer. Evidently the small fraction of active polymer chains formed from the initiator is susceptible to a very effective chain termination process.

It is reported that methyl acrylate, allyl acetate, vinyl acetate and dimethyl maleate give only low yields of oligomers with butyllithium under all experimental conditions (31). FURUKAWA and coworkers (32) confirm that vinyl acetate will not polymerize and that n-butyl-vinyl-ether will not either. High polymers can be formed from isopropyl acrylate (39) in toluene at −70° and from t-butyl acrylate (65). The reported failure of methyl acrylate and butyl acrylate to yield high polymers could reflect a genuine difference in behaviour connected with the side group or could simply result from failure to choose the most favourable conditions for polymerization. Vinyl acetate can be polymerized by lithium metal (49) but co-polymerization experiments suggest that the polymer is formed by a radical mechanism.

Acrylonitrile and butyllithium give good yields of high molecular weight polymer in petroleum ether (31) or in toluene (64). In petroleum ether the polymerization seems to proceed faster and to give higher molecular weight products at −50° than at room temperature. At −15° it was reported that the molecular weight of the polymer formed was independent of the initiator concentration. As no account was taken of the widely different conversions obtained, this result is misleading. The initiator efficiency was in fact the same at all initiator concentrations which suggests that at complete conversion, the molecular weight would be inversely proportional to initiator concentration. Not all initiator molecules however lead to high polymer and some other products must be formed. Only in tetrahydrofuran does the system give a more reasonable balance between initiator molecules used and high polymer formed. In toluene at −78° the polymerization fails to go to completion except at low monomer concentrations. The polymer is not soluble in the system and a gel is formed. The failure to polymerize completely may thus be due to monomer diffusion being slow at high conversions or to a progressive termination reaction. The molecular weight distribution of the polymer formed consists of a sharp peak superimposed on a broader distribution, but the overall $\overline{M}_w/\overline{M}_n$ ratio is about 1.2−1.3. The number-average degree of polymerization is 20−150 times the initial monomer to initiator ratio which again indicates side reactions of butyllithium with the monomer or failure to isolate low molecular weight material. The polymers formed at −78° are colourless, but as the polymerization temperature is raised, they become increasingly yellow, which suggests some cyanoethylation is taking place.

Vinyl chloride can be polymerized by butyllithium, ethyllithium or the complex of butyllithium with triethyl aluminum in hydrocarbon solvents (33). The polymerizations show rather peculiar behaviour for

an anionic system in that air seems to have little effect and polymerization will not proceed in the presence of tetrahydrofuran. With butyllithium, the molecular weight of the polymer isolated seems to be
largely insensitive to initiator concentration or conversion but is proportional to the initial monomer concentration used. The observed
degrees of polymerization (100—300) are much larger than expected
from the monomer-initiator ratio. Once again other products, either
oligomers or those from other reactions of butyllithium with the monomer
must be considered. The conversions observed were low and from the
data presented, polymerization probably stops before all the monomer is
consumed. It is evident that the polymerization process is complex and
perhaps not anionic in nature. The authors in fact suggest from copolymerization studies that it proceeds cationically. This is not too
plausible a suggestion for vinyl chloride, and the exact mechanism is
in doubt.

Vinylidene chloride also gives some high polymer with butyllithium
in hexane (52). Yields are, however, very low. The conversion attained
can be increased by incremental addition of the initiator. This procedure
also increases the molecular weight of the polymer obtained. It is observed
that higher molecular weight polymer is obtained at lower initiator
concentrations, so incremental addition probably produces a set of
independent, self-terminating polymerizations. Other reactions undoubtedly occur since the chlorine content of the polymer is reported to
decrease progressively below the theoretical value as the initiator
concentration is increased. Most of the butyllithium must form products
other than the high polymer reported, for in some cases only one to two
molecules of monomer were consumed per butyllithium molecule added.

V. Polymerization in polar solvents
a) Non-polar monomers

Small amounts of polar solvents such as tetrahydrofuran, ether,
dioxane and triethylamine have been shown to break down the association of organo-lithium compounds in non-polar solvents, and to greatly
increase the rate of chain initiation. In polar solvents, therefore, one
expects rapid initiation and a polymerization rate essentially determined
by the rate of chain propagation of solvated ion-pairs.

Some deviations from this pattern can be expected under certain
conditions. If the concentration of active species is sufficiently low and
the dielectric constant of the solvent is relatively high, enough free
anions can be present to make a large contribution to the reaction. With
solvents which have a lower dielectric constant and are less basic, a
free ion contribution is not likely to be important at practicable concentrations, but at high concentrations of growing species some ion-pair

association is possible. In addition, both lithium alkyls and the polymer ions can react with ethers at an appreciable rate particularly at room temperature. Most polymer carbanions isomerize, particularly in solvents of relatively high dielectric constant. This process is often slow compared with polymerization rates, but not always so. A check for its occurrence should always be made for it may invalidate the rate constant measurements. In tetrahydrofuran reaction rates are often very fast and a flow method must be used. SZWARC (107) has determined a number of propagation constants in this way. The experiments generally refer to sodium as counter-ion.

With polystyryllithium in tetrahydrofuran (4) the effect of ion-pair dissociation is very noticeable. At growing chain concentrations about 10^{-4} molar the rate is about half order in polystyryllithium although the ion-pair is still the major species. This comes about because despite the low concentration of free anions, they are immensely more reactive. k_p for the free ions at $25°$ is found to be 65,000 litre/mole sec whereas the tentative value for the ion-pair is about 200 litre/mole sec. According to Szwarc variation of the counter ion from Li^+ to Cs^+ decreases the reactivity of the ion-pair in regular progression to about half its original value. BAGDASAR'YAN (77) reports at $2°$ that the ion-pair rate constants are all equal at 140 litre/mole sec with the exception of that for polystyrylsodium which is 180 litre/mole sec. His measurements were made in a restricted range at rather high concentrations where the contribution of the free anion is naturally rather small. They may still be liable to a small correction for the free ion reaction. None of the data at present is accurate enough to properly distinguish differences in rate constants of about 50%. The differences in ion-pair dissociation constant are large and easily distinguishable. They decrease rapidly from the lithium compound to that of cesium. The differences are attributable to a decrease in solvation as the cation size increases (4).

Butadiene and isoprene have also been studied in tetrahydrofuran (72). At $0°$ the rates are close to first order in polyisoprenyl or polybutadienyllithium concentration which indicates that the rate constant for the ion-pair is being measured in the concentration range studied ($> 10^{-3}$ molar). The rate constants at $30°$, k_p (butadiene) $= 1.8$ litre/mole sec, k_p (isoprene) $= 0.13$ litre/mole sec, are appreciably lower than for styrene. For butadiene at $-39°$ a k_p value of 6×10^{-2} litre/mole sec can be derived from the results of SPIRIN (98). This value checks well with that extrapolated from Morton's data. The observed propagation constant for isoprene is rather low and is in fact equal to that of the monoetherate in solvent mixtures of appreciably lower dielectric constant. At room temperature there is evidence for isomerization of polyisoprenyllithium in tetrahydrofuran which becomes particularly marked as the monomer concentration reaches low values (48). The isomerized form still seems to be reactive towards monomer to re-form the normal anion but probably reacts at a decreased rate. The isomerization may have some effect on the measured rate constant but it probably is not large at moderate conversions.

In ether the propagation rate for polyisoprenyllithium (0.03 litre per mole sec) is even lower than in tetrahydrofuran (93). The propagation rate is first order in polyisoprenyllithium in the concentration range 10^{-4} to 5×10^{-3} molar but above this range the observed order decreases. The authors assumed that the polyisoprenyllithium was extensively dissociated to free ions and that the rate constant derived refers to the free carbanion. From the variation of rate with concentration of polyisoprenyllithium a dissociation constant of 2.5×10^{-2} was derived for the dissociation to free ions. This value is too high by a factor of about 10^8 over that expected in a solvent of dielectric constant about four. It seems more reasonable to assume that the major species present is the

Table 1. k_p (litre/mole sec)[1]

Polymer	Solvent	Temperature °C	k_p	A (litre/mole sec.)	E (kcal/mole)	Ref.
Polyisoprenyl-lithium	heptane	20	0.65	—	—	95, 96
Polyisoprenyl-lithium	hexane	30	4.7	$3.4 \cdot 10^3$	4.1	73, 74
Polyisoprenyl-lithium	cyclohexane	20	>20	—	—	126
Polyisoprenyl-lithium · THF	hexane	30	0.14	—	—	73
Polyisoprenyl-lithium	ether	20	0.03	—	—	93
Polyisoprenyl-lithium	tetrahydrofuran	30	0.13	$1.5 \cdot 10^4$	6.8	72
Polyisoprenyl-lithium	triethylamine	(±30 ?)	0.17	—	10.7	98, 101
Polystyryl-lithium	benzene	30	>30	—	<14	124
Polystyryl-lithium · 2 THF	benzene	20	0.6	$3.3 \cdot 10^7$	10.4	14
Polystyryl-lithium	triethylamine	—34	0.025	—	—	98, 101
Polystyryl-lithium	tetrahydrofuran	25	200	—	—	4
Polystyryl-lithium	tetrahydrofuran	2	140	—	—	77
Polybutadienyl-lithium	tetrahydrofuran	30	1.8	$4.3 \cdot 10^4$	6.1	72

[1] MORTON [Polymer Preprints 5, 1092 (1964)] has reported for polystyryl-lithium in benzene, $k_p = 1.1 \cdot 10^4$ exp. (—3800/RT). $k_p (30°) = 17.3$ litre/mole sec.

ion-pair and that the rate constant observed refers to this species. The decrease in order at high concentrations of polyisoprenyllithium could then be attributed to a small amount of self-association. Ether is much less basic than tetrahydrofuran and there could be some competition in this solvent between etherate formation and self association.

SPIRIN and co-workers (98, 101) have reported the polymerization of isoprene, butadiene and styrene in triethylamine. With styrene at $-34°$ the rate is first order in polystyryllithium in the concentration range 0.4 to 3.2×10^{-2} molar. The observed propagation constant, 2.5×10^{-2} litre/mole sec should refer to the ion-pair reaction. With isoprene and butadiene in triethylamine the plots of rate against polyalkenyllithium concentration curve somewhat at higher concentrations. This could be due to some self-association because the concentrations used were rather high. Some uncertainty applies to the data on polyisoprene polymerization since the experiments may have been carried out at $+30°$ (101) or at $-30°$ (98).

All the absolute values of k_p reported in the literature are collected in Table 1. In all cases the figures should refer to the reaction of the ion-pair with monomer. The results are too fragmentary and in some cases of uncertain accuracy for a detailed discussion of the effect of environment on reactivity. A few points are clear. The reactivity of the relatively unsolvated ion-pair in hydrocarbon solvents is relatively large and may even be comparable with that of the solvated ion-pair in tetrahydrofuran despite the large difference in dielectric constant. The reactivity of the ether-solvated ion-pair in solvents of lower dielectric constant is lower than either. The first effect of etherate formation is to decrease the reactivity of the ion-pair which can be increased again by an increase of the dielectric constant of the solvent.

b) Polar monomers in polar solvents

With these monomers polymerization again is complicated by the possibility of side reactions and termination reactions. In the butyl-lithium initiated polymerization of methylmethacrylate in tetrahydrofuran at $-78°$, monomer added repeatedly after quite long time intervals will polymerize rapidly (117). This is found to be true even if the solution is warmed to room temperature after each addition of monomer. The possibility remains that some termination occurs and that some reinitiation by lithium methoxide is also important. Similar experiments with fluorenyllithium under the same conditions (40) show that after 30 minutes a large percentage of the polymer chains is reactive towards tritiated acetic acid. Addition of a second batch of monomer at $-78°$ indicates that some chain termination occurs at this temperature but that no new chains are formed by re-initiation. Some other low molecular weight product must be formed however since there is a large discrepancy between the observed degree of polymerization and the monomer/initiator ratio used. With lithium naphthalene as initiator, a much better balance is observed between calculated and experimental molecular weights (22). Side reactions are probably less important with this initiator but still cannot be completely excluded on the evidence available.

t-Butyl vinyl ketone can be polymerized by lithium-biphenyl, butyllithium or even lithium methoxide in tetrahydrofuran to give high polymers (85). The polymerizations do not proceed to complete conversion, although the conversion attained is increased at lower temperatures. This could be a "ceiling temperature" effect as the monomer contains a bulky side-group. Alternatively the results could be due to the presence of a slow termination step which is less effective at lower temperatures. The former explanation has the disadvantage that the thermodynamic parameters derived from "equilibrium" monomer concentrations are not reasonable for the polymerization process. The "equilibrium" monomer concentration also depends on reaction conditions to an extent which is not too plausibly accounted for by changes in the activity coefficient of the monomer. The latter alternative seems more plausible in view of the evidence for termination in similar systems. It has the disadvantage that the likely termination product, lithium methoxide, would clearly re-initiate new chains under the experimental conditions used although it is not clear from the data how effective an initiator lithium methoxide is. Kinetic results at 0° with lithium biphenyl show that the intrinsic viscosity of the polymer increases with conversion as expected of this type of polymerization. With butyllithium as initiator the intrinsic viscosity of the polymer varied little at conversions above 10% which suggests that the polymerization process cannot be described in simple terms of "living" polymerization with a slow termination step.

Methacrylonitrile can be polymerized almost instantaneously at −75° in liquid ammonia with lithium metal as initiator (83, 84). It was suggested that initiation occurs by a rapid electron transfer to monomer followed by a fast anionic reaction. Lithium amide produced in the reaction itself is not the initiator for it is a comparatively slow initiator of polymerizations at the temperature used. The polymer ions apparently abstract a proton from ammonia to form lithium amide which then reacts with nitrile groups on the polymer to produce cyclic structures. It is believed that this reaction is slow compared to the polymerization process.

VI. Copolymerization

In systems where chain termination is absent, anionic polymerizations provide in principle an excellent method for the preparation of

block-copolymers (*109*). As the propagation steps in anionic polymerizations are highly discriminatory, the method will fail for monomers of
widely different polarity. The polymethylmethacrylate ion-pair for
instance will not add styrene (*41*). With polar monomers the copolymerization will be complicated by the side reactions described earlier. Most
of the work in this field has been carried out in tetrahydrofuran with
sodium based initiators and so is outside the scope of this review. The
lithium based systems would be of value where carbanion isomerization is
known to be a complicating factor. The polymerization could be carried
out in hydrocarbon solvents in the presence of trace amounts of polar
materials.

In more conventional copolymerization, where both monomers are
present at the start of reaction, standard analysis methods are available
which were developed for free radical initiation. It is first necessary to
enquire whether the techniques will be applicable to anionic systems
containing long lived active species. Four propagation steps can be
recognized in the copolymerization of two monomers M_1 and M_2:

$$1) \quad P_1^* + M_1 \xrightarrow{k_{11}} P_1^*$$

$$2) \quad P_1^* + M_2 \xrightarrow{k_{12}} P_2^*$$

$$3) \quad P_2^* + M_2 \xrightarrow{k_{22}} P_2^*$$

$$4) \quad P_2^* + M_1 \xrightarrow{k_{21}} P_1^*$$

where P_x^* represents an active polymer chain having an x unit at the
active centre. The concentration of P_x^* will be the concentration of
the free ion-pair in solvents where the active species are associated. The
system is usually analyzed in terms of two reactivity ratios, $r_1 \left(= \dfrac{k_{11}}{k_{12}} \right)$
and $r_2 \left(= \dfrac{k_{22}}{k_{21}} \right)$ and the reactivity ratios are determined from the copolymer composition by the equation

$$\frac{d[M_1]}{d[M_2]} = \frac{[M_1]\{r_1[M_1] + [M_2]\}}{[M_2]\{r_2[M_2] + [M_1]\}}$$

where $\dfrac{d[M_1]}{d[M_2]}$ represents the relative rate of incorporation of the two
monomers in the copolymer and $[M_1]$ and $[M_2]$ are the concentrations
of the two monomers in the mixture. The equation is usually derived by
assuming that a stationary concentration of one of the active centres is
rapidly established. In anionic systems even when chain initiation is
complete we can only assume that the *total* concentration of active
species is constant. However the only real requirement is that an equilibrium between the two types of active centre be established. This

requires that the rates of the two cross-propagation reactions should equalize rapidly and will occur for any system in which the block lengths are not exceptionally long. The copolymer composition equation should therefore be a reasonable approximation even where chain initiation is not complete if the copolymer composition is measured at the point where the polymer molecular weight is sufficiently large that most polymer molecules contain several alternations of each monomer in the chain. If one of the cross-propagation constants is zero (styrene, methyl-methacrylate) alternation cannot occur and the copolymer composition equation takes a special form as shown by O'DRISCOLL (78). As the monomer concentrations change continuously except under special conditions $(r_1 = r_2 = 1)$ it is customary to extrapolate the copolymer composition to zero conversion. This is a potentially dangerous procedure for anionic systems for it is equivalent to extrapolation to very low molecular weight and the conditions expressed above may not hold. In addition, with low molecular weight polymer preferential initiation to one monomer may influence the copolymer composition (41). The two requirements are obviously mutually contradictory and must limit the applicability of the copolymer composition equation for determination of reactivity ratios in anionic systems. In practice the copolymer compositions are often measured at conversions which are not less than 10—20%. This could be a reasonable compromise with many systems, but it is probable that the reactivity ratios in the literature can only be considered as qualitative estimates.

In hydrocarbon solvents it is known that most of the growing chains are associated and it is necessary to enquire what effect this has on the copolymerization mechanism. The reactivity ratios measured from copolymer composition are unaffected because they refer to a common ion-pair. The equilibrium constants for association cancel and the reactivity ratios measured give a true measure of the relative propagation constants of the two monomers. No assessment can be made of the real reactivity of two types of active chain with the same monomer, however. In this case the observed rates are a function of the relative reactivities of the free ion-pairs and also of the relative extents of association. For example in hydrocarbon solvents polystyryllithium reacts with butadiene much more rapidly than does polybutadienyllithium. Until we know the two equilibrium constants for self-association we cannot find out if the increased rate is due to greater intrinsic reactivity or to a higher concentration of free polystyryllithium. In polar solvents or in hydrocarbon solvents in the presence of small amounts of ethers, these difficulties do not arise as self-association is no longer important.

As might be expected the reactivity ratios are strongly dependent on the solvent system. The amount of styrene in the initial copolymer formed from a 60:40 mixture of styrene and isoprene with butyllithium is reported to increase from 18% in benzene to 60% in triethylamine, to 68% in diethylether and reaches 80% in tetrahydrofuran (*50*). Only 4% of diethylether in benzene increases the copolymer styrene content almost to that value observed in diethylether itself. In the copolymerization of styrene and butadiene, a similar effect of ether is observed (*56*), and tetrahydrofuran is even more efficient (*60*) since the copolymer composition is reported to reach the limiting value when only one molecule of tetrahydrofuran is present for each polymer chain. It appears that the copolymer composition is mainly determined by the nature of the solvation around the ion-pair and is not greatly dependent on the bulk properties of the medium.

Interesting effects are observed in the copolymerization of butadiene-styrene, isoprene-styrene or isoprene-butadiene in hydrocarbon solvents. In each case, the copolymerization rate during the first half of the reaction is essentially that of the monomer which homopolymerizes more slowly (*54, 57, 60, 87, 102*). The initial copolymer formed is enriched in this monomer. In copolymerization with styrene, in the latter half of the reaction, the colourless solution takes on the colour associated with polystyryllithium and the rate increases to that of the homopolymerization of styrene. KOROTKOV (*87*) who first observed this effect, suggested that it is due to a preferential solvation of the growing chain end by one of the two monomers, and hence the copolymerization rate becomes equal to that of the homopolymerization of this monomer. This viewpoint was clarified by the suggestion that the "solvation" should be considered as the formation of a relatively stable complex of the active centre with monomer (*57*). The complex would be similar to the etherates which are known to be formed with organo-lithium compounds. Chain propagation would then occur by the interaction of a further monomer molecule with the complex. It does not seem necessary to postulate complex structures of this type to explain the results. It is misleading to compare copolymerization rates with homopolymerization rates for the former rate is dependent on four different propagation rates (*13*) and the magnitude of the "cross-over" constants, k_{12} and k_{21}, can be very important (*78*). Only with a knowledge of all four propagation constants can the mechanism be elucidated.

The butadiene-styrene system alone has received the detailed study required to give a clearer picture of the mechanism. The results should, however, be similar for the other systems. The two homopolymerization rates are easily measured. The exchange rate between two active centres can be measured by forming a solution of polybutadienyllithium or of polystyryllithium and allowing it to react with the other monomer. It is convenient to measure the rate spectroscopically from the rate of

appearance or disappearance of polybutadienyllithium or polystyryllithium. In this way, Morton and Ells (71) find r_1 (butadiene) = 3 and
r_2 (styrene) = 0.06 in benzene solution at 29°. The results show that polystyryllithium reacts much too rapidly with butadiene for an accurate
determination of k_{21}, and that the high rate of this cross-propagation
reaction is responsible for the "inversion" phenomenon in the copolymerization rate.

Similar results are obtained in cyclohexane at 40° ($r_1 = 26; r_2 < 0.04$).
Under these conditions the reaction of polystyryllithium with butadiene
occurs virtually instantaneously and with comparable concentrations of
both monomers present the homopolymerization rate of styrene can be
neglected (46). Hence:

$$-\frac{d[M_1]}{dt} = k_{11}[P_1^-][M_1] + k_{21}[P_2^-][M_1]$$

$$= k_{11}[P_1^-][M_1] + k_{12}[P_1^-][M_2]$$

$$-\frac{d[M_2]}{dt} = k_{12}[P_1^-][M_2]$$

$$-\frac{d[M_1 + M_2]}{dt} = k_{11}[P_1^-][M_1] + 2k_{12}[P_1^-][M_2]$$

where $[P_1^-]$ and $[P_2^-]$ are the concentrations of *free* polybutadienyllithium and polystyryllithium, respectively. With $k_{12}/k_{11} = 0.038$,
then at equimolar monomer concentration the copolymerization rate
should be 8% higher than that of the homopolymerization rate of
butadiene and the initial copolymer should contain about 4 mole percent
of styrene. The small increase in rate is close to the limits of error of
measurement and is hard to detect but copolymer composition measurements confirm the predicted value. k_{12} is measured in the absence of
butadiene monomer, whereas the copolymer composition is naturally
determined from experiments where both monomers are present. The
agreement between calculated and experimental compositions suggests
that k_{12} is not affected by the presence of butadiene. This point can be
confirmed independently. The reaction of polybutadienyllithium with
styrene can be studied in the presence or absence of butadiene. No
difference in the rates can be detected within the experimental error (46).
The theory of preferential absorption (or complex formation) around the
growing centre cannot be valid for the major active species present.

In hydrocarbons, butadiene is the more reactive monomer with
either polybutadienyllithium or polystyryllithium. Association of either
active species has no effect in this comparison. The polystyryl anion
seems to be more reactive with either monomer but we do not know if
this effect is real or caused by a greater equilibrium concentration of free
polystyryllithium. It may be significant that this compound exists mainly

as a dimeric species whereas polybutadienyllithium is hexameric. For this reason alone large differences in the apparent relative reactivity would be expected between non-polar and polar solvents.

Copolymerizations initiated by lithium metal should give the same product as produced from lithium alkyls. Usually the radical ends produced by electron transfer initiation have so short a lifetime they can have no influence on the copolymerization. This is true for instance in the copolymerization of isoprene and styrene (50). The product is identical if initiated by lithium metal or by butyllithium. With the styrene-methylmethacrylate system, however, differences are observed (79, 80, 82). Whereas the butyllithium initiated copolymer contains no styrene at low conversions, the one initiated by lithium metal has a high styrene content if the reaction is carried out in bulk and a moderate one even in tetrahydrofuran. These facts led O'DRISCOLL and TOBOLSKY (80) to suggest that initiation with lithium occurs by electron exchange and that in this case the radical ends are sufficiently long-lived to produce simultaneous radical and anionic reactions at opposite ends of the chain. Only in certain rather exceptional circumstances would the free radical reaction be of importance. Some of the conditions required have been discussed by TOBOLSKY and HARTLEY (111). The anionic reaction should be slow. This is normally true for lithium based catalysts in hydrocarbon solvents. No evidence of appreciable radical participation is observed for initiation by sodium and potassium. The monomers should show a fast radical reaction. If styrene is replaced by isoprene, no isoprene is found in the copolymer for isoprene polymerizes slowly by free radical initiation. Most important of all, initiation should be slow to produce a low steady concentration of radical-anions. An initiator which produces an almost instantaneous and complete electron transfer to monomer produces a high radical concentration which will ensure their rapid mutual termination.

It has been suggested (76) that the styrene is incorporated only in the initiation reaction and that propagation is purely anionic. At low conversions the styrene found in the copolymer in the butyllithium-initiated copolymerization at high styrene/methylmethacrylate feed ratios is probably introduced in the initiation step (41). For the lithium initiated reaction in bulk, the degree of polymerization is quite high at low conversions ($\sim$ 7000) (35) and as the initiation reaction could only produce a few styrene units per chain it cannot be responsible for the large amounts of styrene found. It is surprising that the measured degree of polymerization is so high and changes little from 1 to 10% conversion. It seems likely that much of the polymerization is produced in a radical reaction and that anionic propagation is suppressed by termination reactions, either those naturally occurring with methylmethacrylate or

Table 2. *Reactivity Ratios in copolymerization initiated by lithium compounds*

M_1	M_2	solvent	initiator	temperature °C	r_1	r_2	ref.
styrene	butadiene	toluene	EtLi	25	0.1	12.5	98
styrene	butadiene	benzene	BuLi	30—50	0.05	15	54
styrene	butadiene	heptane	BuLi	30	~0	7	60
styrene	butadiene	benzene	BuLi	29	0.06	3	71*
styrene	butadiene	cyclo-hexane	BuLi	40	<0.04	26	46*
styrene	butadiene	ether	BuLi	30	0.11	1.78	56
styrene	butadiene	THF	EtLi	—35	8	0.2	98
styrene	butadiene	THF	BuLi	30	0.77	1.03	67
styrene	isoprene	toluene	EtLi	27	0.25	9.5	98
styrene	isoprene	benzene	BuLi	30	0.14	7	57
styrene	isoprene	THF	EtLi	27	9	0.1	98
styrene	isoprene	THF	EtLi	—35	40	~0	98
isoprene	butadiene	hexane	BuLi	50	0.47	3.4	87
isoprene	1,3 penta-diene	hexane	BuLi	50	17	0.06	45
styrene	p-methoxy-styrene	THF	Li	0	2.9	0.23	110
styrene	p-methoxy-styrene	toluene	BuLi	0	10.9	0.05	110
styrene	p-methyl-styrene	THF	Li	0	1.3	0.9	110
styrene	p-methyl-styrene	toluene	BuLi	0	2.5	0.26	110
p-methyl-styrene	p-methoxy-styrene	THF	Li	0	1.93	0.72	110
methylmeth-acrylate	acrylo-nitrile	bulk	BuLi	—8	0.39	7	128
styrene	acrylo-nitrile	iso-octane or ether	BuLi	—12	0.20	14	128

* From rate measurements; all other data from copolymer composition.

caused by trace impurities to which the anionic reactions are particularly susceptible.

A number of reactivity ratios have been determined from initial copolymer composition data. These are recorded in Table 2. In view of the difficulties associated with their determination and uncertainty whether the copolymer composition equation is accurate under all conditions, they should be considered as of unknown accuracy.

VII. Stereospecificity of the propagation step
a) Diene polymerization

The polymer formed from isoprene with lithium has a predominantly cis-1,4 structure whereas that from the other alkali metals has a more mixed microstructure (*103*). The predominantly cis-1,4 configuration is retained on dilution with hydrocarbon solvents and with lithium alkyls, but in polar solvents 3,4 linkages predominate (*44, 45*). With butadiene

(*100, 101, 102*), lithium and its compounds produce a largely 1,4 structure in bulk or in hydrocarbons. It is, however, of mixed cis and trans content (*28*). On addition of polar substances higher amounts of 1,2 structures are introduced (*53, 58*) into the polymer.

The microstructure of polyisoprene prepared in a variety of solvents and solvent mixtures (*113*) has been determined. Various ethers and sulphides vary in their ability to reduce the 1,4 content of the polymer. The most effective ether was tetrahydrofuran. The presence of only two molecules per active chain was reported to reduce the 1,4 content to that observed in the pure ether. More recent investigations have failed to confirm that the requirement is as low as this (*74, 126*) but relatively small amounts of tetrahydrofuran do markedly decrease the cis-1,4 content and increase the 3,4 content. Similar results have been obtained for butadiene (*60*) with respect to 1,4 and 1,2 structures.

The microstructure of the polymer varies little with changing reaction conditions (*68, 104*). The effect of temperature is generally small and the alkali metals or their alkyls normally give the same product. Significant differences in microstructure have been noted between potassium and its alkyls (*104*) and between two different cesium compounds (*88*) but these effects are not general and their cause is obscure. A more difficult problem exists in that there is poor agreement between the microstructures reported by different authors for a particular initiator and solvent. Tables 3 and 4 include some of the data given for polyisoprene and polybutadiene. Standard infra-red methods were used for the analysis except

Table 3. *Microstructure of polyisoprene*

(Li)				(Na)				(K)				ref.
cis	trans	3,4	1,2	cis	trans	3,4	1,2	cis	trans	3,4	1,2	
in bulk, hydrocarbons or benzene												
91	—	9	—	—	12	77	11	—	—	—	—	*68*
93	—	7	—	—	46	45	9	—	55	36	9	*113*
94	—	6	—	3	40	50	7	20	40	32	7*	*104*
94	—	6	—	—	43	51	6	—	52	40	8	*28*
80	15	5	—	29	29	42	—	—	—	—	—	*126*
in diethylether												
—	28	65	7	—	—	—	—	—	—	—	—	*68*
—	49	46	5	—	33	55	12	—	51	39	10	*113*
6	29	60	5	—	14	76	10	4	27	63	6	*104*
—	—	—	—	—	7	74	19	—	—	—	—	*16*
in tetrahydrofuran												
—	3	65	32	—	81	9	10	—	—	—	—	*68*
—	33	51	16	—	33	54	13	—	48	36	16	*113*
—	—	74	26	—	—	82	18	—	—	—	—	*126*

* Microstructure with potassium alkyl. With potassium metal: 0, 52, 40, 7.

Table 4. *Microstructure of polybutadiene*

(Li)			(Na)			(K)			(Rb)			(Cs)			ref.
cis	trans	1,2	cis	trans	1,2	cis	trans	1,2	cis	trans	1,2	cis	trans	1,2	

in bulk, hydrocarbons or benzene

cis	trans	1,2	cis	trans	1,2	cis	trans	1,2	cis	trans	1,2	cis	trans	1,2	ref.
47	43	10	—	—	—	—	—	—	—	—	—	—	—	—	61
43	49	8	23	45	32	15	49	36	—	—	—	—	—	—	1
35	52	13	10	25	65	15	40	45	7	31	62	6	35	59	28
—	—	—	—	—	—	—	—	—	—	—	—	—	43	57	88

in tetrahydrofuran

cis	trans	1,2	cis	trans	1,2	cis	trans	1,2	cis	trans	1,2	cis	trans	1,2	ref.
—	9	91	—	12	88	—	15	85	—	—	—	—	—	—	1
—	4	96	—	9	91	—	18	83	—	25	75	—	25	75*	88

* With cesium naphthalene; 0, 44, 56, with cesium butadiene.

for references (*16*) and (*126*) where an N. M. R. method was used. Results have been averaged if there was a small difference between the alkali metal and its alkyls otherwise they are noted. They generally refer to 0° or room temperature with the exception of reference (*1*) which applies to −50°. The earlier results of MORITA and TOBOLSKY (*68*) are quoted as well as those of TOBOLSKY and ROGERS (*113*). The authors considered the later results to be more accurate but as the earlier ones sometimes agree better with other estimates they are included in the table.

The results for butadiene are reasonable consistent. The infra-red bands which are associated with cis-1,4, trans-1,4 and 1,2 structures are well separated and there is about a ten to fifteen percent variation between various suggested analysis procedures (*61*).

The standard infra-red methods are clearly inadequate for poly-isoprene. The determination of relative amounts of cis and trans structures in particular is subject to large errors (*113*) as the infra-red bands are virtually coincident and differ only in band shape and intensity. The N. M. R. method of analysis (*16*) is more satisfactory, particularly for determination of the cis/trans ratio[1]. As can be seen from the table the results are often widely different from those obtained by infra-red analysis.

Owing to uncertainties in the analysis methods, only general trends seem to warrant discussion. With butadiene, the 1,2/1,4 ratio increases in the series Li to Cs. In tetrahydrofuran the polymer is even more rich in 1,2 structures and varies little with the alkali metal or its compounds. With isoprene, only lithium and its compounds give a highly cis-1,4 polymer in hydrocarbons. Increasing amounts of the 3,4 structure occur with the other alkali metals. The amount of 1,4 polymer and its internal distribution is in doubt. In ethers the 3,4 polymer is the major constituent with all initiators. The N. M. R. data suggest the microstructure

---

[1] Two recently developed infra-red methods (*74, 94*) give better agreement with N. M. R. data for the cis/trans ratio.

is about 80% 3,4 and 20% 1,2. Infra-red studies suggest an appreciable trans-1,4 content.

It is generally accepted that the microstructure is dependent on the degree of ionic character of the carbon-metal bond (*99, 112*). It is usually considered that in hydrocarbons the carbon-lithium bond is largely covalent, whereas the bonds with other alkali metals are increasingly ionic in the series Li, Na, K, Rb, Cs. In good solvating solvents and even in the presence of small amounts of them, the ionic character is increased. The maximum solvent effect should occur with lithium compounds of course, and with the heavier alkali metals polar solvents should have little effect, as is in fact observed. With the ionic reaction small variations could be attributed to specific solvation effects, and in extreme cases some changes caused by reaction of the free anion might be important.

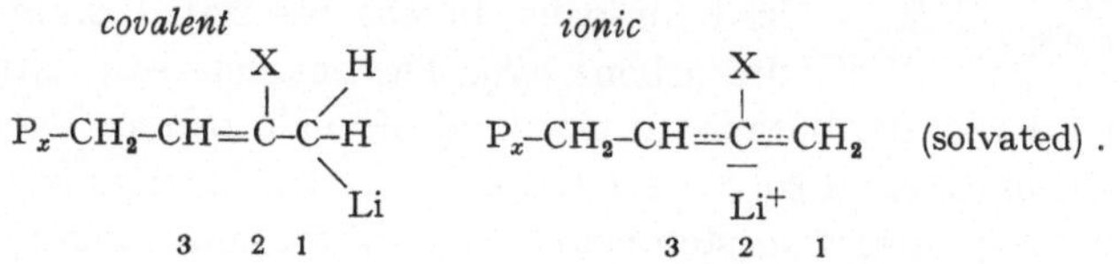

A more detailed description of the mechanism of isoprene polymerization by lithium compounds has been given (*99, 104*). The polyisoprenyllithium first complexes with isoprene in the cis-form. The complex subsequently rearranges to form a transition state in the form of a six-membered ring.

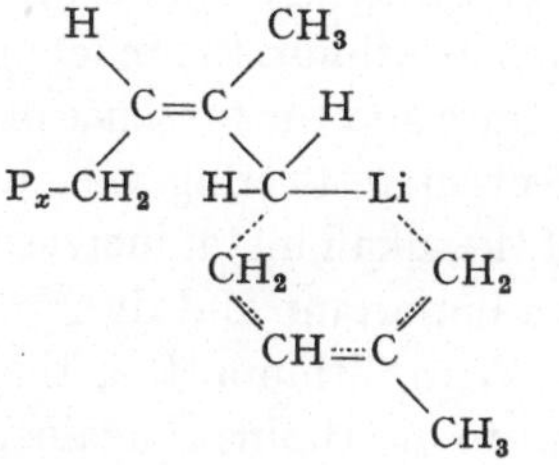

The configuration of each monomer unit in the chain is thus fixed at its point of entry. A similar transition state was suggested by SZWARC (*106*) where the incoming monomer molecule approaches above the plane of the last unit of the polymer chain. Only isoprene in the cis-form would be capable of reaction. 1,2 addition is also possible with this mechanism but its occurrence is considered less likely owing to the greater electron availability at the terminal carbon atoms, and to the lower stability of the olefin formed.

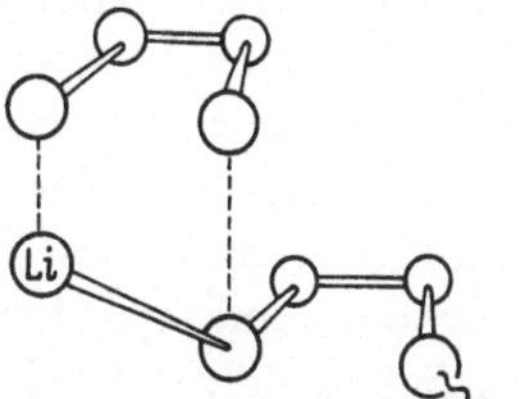

Fig. 8

The ion pair form, postulated for lithium compounds in THF or for the higher alkali metals even in hydrocarbons would react preferentially at the 3-carbon atom as the electron availability is higher at this point (essentially it is a secondary carbanion) leading to 1,2 or 3,4 addition. Addition at the 1-position is less likely and would necessitate the monomer approaching in the trans-form as shown in the diagram.

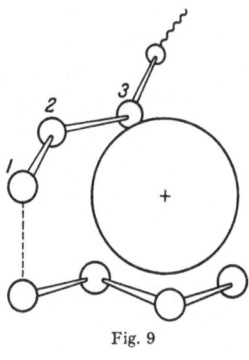

Fig. 9

Mechanisms of the above type are very plausible but two points should be considered. Firstly, all these transition states are equally plausible for butadiene and isoprene whereas butadiene gives a mixed cis-trans product with lithium alkyls in hydrocarbons. Secondly, it is not certain that these carbon-lithium bonds are essentially covalent in hydrocarbons. There is evidence that the lithium compounds of conjugated monomers still exist as charge delocalized ion-pairs in the associated state in hydrocarbons (48). The characteristic ultra-violet absorption band attributable to this kind of anion pair persists almost unchanged in different solvents and alkali metals. The monomeric form active in the propagation step could possibly contain a more covalent carbon-lithium bond but we cannot be sure of this.

It is clear that the only real characteristic of the lithium initiated systems in hydrocarbons is 1,4 addition. The only important difference between isoprene and butadiene is the presence of the methyl group on the monomer and chain end with isoprene. It seems reasonable to suppose that with a small ion such as lithium the reactants must approach rather closely in the transition state and the presence of the methyl groups could have an important effect in restricting the mode of approach of the monomer. As the size of the alkali metal increases the presence of methyl groups will no longer be important and discrimination between cis and trans would decrease. With lithium too, the closeness of approach required would also tend to discriminate against addition at carbon 3 so 1,2 or 3,4 addition would not be favored. Much more space is available for attack at the 1-position. If an ionic model is used throughout, the approaching monomer molecule "pulls off" the counter ion, rather than inserting itself between the carbon-metal bond, the closest approach being at the carbon centres to be bonded. The angle between the two essentially planar reacting species increases as the effective size of the counter-ion increases (effective in the sense of either intrinsic size or solvation) and repulsions between CH_2 and CH_3 groups of monomer and chain-end decrease. This mechanism appears to be as plausible as the generally accepted one but it should be remembered that all theories of the intimate geometry of the transition state are highly speculative[1].

---

[1] With 1,4-disubstituted dienes, more complex stereochemistry is possible in the polymers. In addition to the possibility of geometrical isomerism, there are two types of completely assymetric centre in the molecule and tritactic polymers can

b) Methylmethacrylate and other acrylates

Polymerization of methylmethacrylate by organo-lithium reagents yields polymers of different microstructure according to the solvent used (*29, 36*). Highly isotactic polymer is formed in solvents such as toluene and highly syndiotactic polymer in polar solvents such as dimethoxy-ethane. Stereoblock polymers are formed in toluene containing small amounts of ethers. [The syndiotactic and isotactic assignments are reversed in the preliminary paper and are corrected later (*105*).] More detailed data on changes of microstructure with solvent and for lithium sodium, and potassium alkyls have been given by BRAUN and co-workers (*6*). The results fall roughly into three groups. In the first, which includes all the alkyls in toluene, there is a preference for isotactic placements which decreases in magnitude in the series Li > Na > K. The second group includes the lithium alkyl in dimethoxyethane and pyridine and the sodium alkyl in dimethoxyethane where the polymer is predominantly syndiotactic. The third group consists of the potassium alkyl in di-methoxyethane and pyridine and the sodium alkyl in pyridine. In the latter group the stereosequence distribution in the polymer follows the simple (Bernoullian) statistics observed for free radical polymerization. One single parameter is adequate to describe the composition as expected where the probability of a monomer adding in a specific configuration is only dependent on the configuration of the last unit on the chain (*115*). It is possible that in the third group polymerization proceeds via free ions. The sequence distributions in the first two groups do not obey simple statistics but those which could be explained by penultimate effects.

Attempts to explain the stereoregulating action of lithium alkyls in non-polar solvents have in fact been made in terms of interactions of the active end with preceding monomer units. GLUSKER (*37*) suggested that isotactic addition occurs because the lithium counter-ion complexes with the carbonyl oxygen of both ultimate and penultimate ester groups on the polymer. Displacement of the penultimate ester group from the complex by that of the entering monomer unit forces it to add in a specific direction. The presence of Lewis bases would of course destroy this effect and hence the high degree of isotacticity. CRAM and KOPECKY (*20*) postu-late complete enolization of the chain end, and that reaction occurs through a semirigid cyclic structure formed by attack of the alkoxide ion on the penultimate monomer unit.

be formed. NATTA has described the preparation of di-iso-*trans*-tactic polymers from alkylsorbates and butyl-β-styrylacrylic acid using optically active lithium alkyls or inactive alkyls with optically active menthylethylether at low tem-peratures in toluene. The polymers were optically active [NATTA, G., M. FARINA and M. DONATI: Makromol. Chem. **43**, 251 (1961)].

Monomer attack occurs from below the ring, as this is the point of least steric hindrance, leading to an isotactic placement. A modification of this mechanism (39) assumes that the mode of attack is not forced by steric reasons but because of the interaction of the carbonyl group of the incoming monomer with the lithium ion. Stereospecific polymerization of acrylates would require this modification. BAWN and LEDWITH (2) present a more detailed description of the transition state to be expected where secondary bonding to the carbonyl group of the penultimate monomer unit is supposed. Retention of configuration will occur if the incoming monomer always presents the same conformation towards the lithium atom. The approach in which the α-methyl groups of the carbanion and monomer are trans was considered to be the most favourable on steric grounds.

CRAM and KOPECKY also suggested a transition state for the polymerization carried by free ions. It is based on the preferred conformation being that which most exposes the active centre to attack and the supposition that the steric effects of the side groups decrease in the order $P_n > COOCH_3 > CH_3$. This would favour syndiotactic placements.

Actually polymerizations in the second group above give the most highly syndiotactic polymers and the sequence distribution is not the simple Bernoullian one expected from such a mechanism. No attempt seems to have been made to formulate a transition state for this type of polymerization. It would obviously lead to difficulties if expressed in terms of penultimate effects as the solvents used are strongly solvating and secondary interaction with penultimate units would not be favoured.

The third group of polymerizations could follow CRAM and KOPECKY's free ion mechanism.

COLEMAN and FOX (18) have pointed out that the non Bernoullian sequence distribution observed in some of these systems can be formed without the hypothesis of penultimate effects. All that is required is that two or more types of active species be present which do not rapidly interconvert. Each can add monomer at its own rate and with its own characteristic regulating effect. No penultimate effect is necessary but the sequence distribution will be non-Bernoullian. This type of mechanism is particularly attractive in the explanation of stereoblock polymer formation in the lithium alkyl systems in toluene with small amounts of ether present. The presence of at least two species of active centres has been inferred from an examination of polymer fractions obtained from butyllithium initiated polymerizations (19) in toluene. The change in molecular weight distribution with time suggests the presence of two or more types of polymer adding monomer at quite different rates. In addition the isotacticity of the fractions is a function of their molecular weight which suggests the species have different stereoregulating action. It was suggested that lithium methoxide formed in the initial stages of reaction is present in the transition state in one of the types of growing end. The addition of lithium methoxide itself however, although it accelerated the reaction, only produced a slightly more isotactic product. Lithium propoxide was found to be much more effective in increasing the isotacticity of the polymer. It does seem, however, that side products formed in the reaction could have some influence on the stereospecificity of the polymer formed.

Some information is available on other acrylates. N,N-disubstituted acrylamides form isotactic polymers with lithium alkyls in hydrocarbons (12). t-Butylacrylate forms crystallizable polymers with lithium-based catalysts in non-polar solvents (65) whereas the methyl, n-butyl, sec-butyl and isobutyl esters do not. Isopropylacrylate also gives isotactic polymer with lithium compounds in non-polar solvents (34). The inability of n-alkylacrylates to form crystallizable polymers may result from a requirement for a branched alkyl group for stereospecific polymerization. On the other hand lack of crystallizability cannot be taken as definite evidence of a lack of stereoregulating influence, as sometimes quite highly regular polymer fails to crystallize. The butyllithium-initiated polymers of methylmethacrylate for instance cannot be crystallized. The presence of a small amount of more random structure appears to inhibit the crystallization process[1].

c) Styrene and its derivatives

It is reported that isotactic polymer is formed from styrene polymerized by butyllithium in hydrocarbon solvents at low temperatures (7, 51). The formation of isotactic polymers is however dependent on the presence of adventitious lithium hydroxide caused by part of the catalyst being destroyed by moisture (125). No isotactic polymer is formed in rigorously purified systems and the microstructure is in fact fairly highly syndiotactic (10). Lithium methoxide or isobutoxide do not induce isotactic polymer formation.

Poly α-methyl styrene is also reported to have a predominantly syndiotactic structure as prepared with butyllithium in cyclohexane (11). SAKURADA and co-workers (90) have suggested that there is the possibility of error in the N. M. R. band assignments and that the polymer may be predominantly isotactic. It is difficult to assess the validity of this claim without details of the crystallization of the supposedly isotactic polymer.

---

[1] The formation of isotactic polymer from allylacrylate suggests that there is no requirement for a branched alkyl group in the side chain [DONATI, M., and M. FARINA: Makromol. Chem. 60, 233 (1963)].

The author is indebted to Dr. D. J. Worsfold and Dr. D. M. Wiles for frequent discussions during the preparation of this manuscript.

Bibliography

1. Basova, R. V., and A. A. Arest-Yakubovich: Doklady Akad. Nauk S.S.S.R. **149**, 1067 (1963).
2. Bawn, C. E. H., and A. Ledwith: Quart. Rev. (Lond.) **16**, 362 (1962).
3. Berkowitz, J., D. A. Bafus, and T. L. Brown: J. Phys. Chem. **65**, 1380 (1961).
4. Bhattacharyya, D. N., C. L. Lee, J. Smid, and M. Szwarc: Polymer **5**, 54 (1964).
5. Bovey, F. A., and G. V. D. Tiers: J. Polymer Sci. **44**, 178 (1960).
6. Braun, D., M. Herner, U. Johnsen, and W. Kern: Makromol. Chem. **51**, 15 (1962).
7. —, W. Betz, and W. Kern: Makromol. Chem. **42**, 89 (1960).
8. Brown, T. L., D. W. Dickerhoof, and D. A. Bafus: J. Am. Chem. Soc. **84**, 1371 (1962).
9. —, and M. T. Rogers: J. Am. Chem. Soc. **79**, 1859 (1957).
10. Brownstein, S., S. Bywater, and D. J. Worsfold: J. Phys. Chem. **66**, 2067 (1962).
11. — — — Makromol. Chem. **48**, 127 (1961).
12. Butler, K., P. R. Thomas, and G. J. Tyler: J. Polymer Sci. **48**, 357 (1960).
13. Bywater, S.: Pure and Appl. Chem. **4**, 319 (1962).
14. —, and D. J. Worsfold: Can. J. Chem. **40**, 1564 (1962).
15. Cheema, Z. K., G. W. Gibson, and J. F. Eastham: J. Am. Chem. Soc. **85**, 3517 (1963).
16. Chen, H. Y.: Anal. Chem. **34**, 1793 (1962).
17. Coates, G. E.: Organo-metallic Compounds (Chapter 1). London: Methuen 1960.
18. Coleman, B. D., and T. G. Fox: J. Polymer Sci. C **4**, 345 (1963).
19. Cottam, B. J., D. M. Wiles, and S. Bywater: Can. J. Chem. **41**, 1905 (1963).
20. Cram, D. J., and K. R. Kopecky: J. Am. Chem. Soc. **81**, 2748 (1959).
21. Dailey, B. P., and C. H. Townes: J. Chem. Phys. **23**, 118 (1955).
22. Datin, A.: Thesis, Sorbonne (1961).
23. Diem, H. E., H. Tucker, and C. F. Gibbs: Rubber Chem. and Technol. **34**, 191 (1961).
24. Dietrich, H.: Acta Cryst. **16**, 681 (1963).
25. East, G. C., P. F. Lynch, and D. Margerison: Polymer **4**, 139 (1963).
26. Evans, A. G., and D. B. George: J. Chem. Soc. **1961**, 4653.
27. —, and N. H. Rees: J. Chem. Soc. **1963**, 6039.
28. Foster, F. C., and J. L. Binder: Advan. Chem. Ser. no. 17, 7 (1957).
29. Fox, T. G., et al.: J. Am. Chem. Soc. **80**, 1768 (1958).
30. Francois, B., V. Sinn, and J. Parrod: J. Polymer Sci. C **4**, 375 (1963).
31. Frankel, M., A. Ottolenghi, M. Albeck, and A. Zilkha: J. Chem. Soc. **1959**, 3858.
32. Furukawa, J., et al.: J. Chem. Soc. Japan (Ind. Chem.) **63**, 640 (1960).
33. — J. Chem. Soc. Japan (Ind. Chem.) **63**, 645 (1960).
34. Garrett, B. S., et al.: J. Am. Chem. Soc. **81**, 1007 (1959).
35. George, D. E., and A. V. Tobolsky: J. Polymer Sci. B **2**, 1 (1964).
36. Glusker, D. L., R. A. Galluccio, and R. A. Evans: J. Am. Chem. Soc. **86**, 187 (1964).
37. —, I. Lysloff, and E. Stiles: J. Polymer Sci. **49**, 315 (1961).
38. —, E. Stiles, and B. Yoncoskie: J. Polymer Sci. **49**, 297 (1961).

39. GOODE, W. E., F. H. OWENS, and W. L. MYERS: J. Polymer Sci. **47**, 75 (1960).
40. GRAHAM, R. K., D. L. DUNKELBERGER, and E. S. COHN: J. Polymer Sci. **42**, 501 (1960).
41. — —, and W. E. GOODE: J. Am. Chem. Soc. **82**, 400 (1960).
42. HARRIES, C.: U. S. Patent 1,058,056 (1913).
43. HEIN, F., and H. SCHRAMM: Z. physik. Chem. A **151**, 234 (1930).
44. HSIEH, H., D. J. KELLEY, and A. V. TOBOLSKY: J. Polymer Sci. **26**, 240 (1957).
45. —, and A. V. TOBOLSKY: J. Polymer Sci. **25**, 245 (1957).
46. JOHNSON, A. F., and D. J. WORSFOLD: Makromol. Chem. (in press).
47. — — J. Polymer Sci. (in press).
48. — —, and S. BYWATER: Can. J. Chem. **42**, 1255 (1964).
49. KELLEY, D. J.: J. Polymer Sci. **59**, s6 (1962).
50. —, and A. V. TOBOLSKY: J. Am. Chem. Soc. **81**, 1597 (1959).
51. KERN, R. J.: Nature (Lond.) **187**, 410 (1960).
52. KONISHI, A.: Bull. Chem. Soc. Japan **35**, 197 (1962).
53. KOROTKOV, A. A.: I.U.P.A.C. Meeting Prague 1957. Angew. Chem. **70**, 85 (1958).
54. —, and N. N. CHESNOKOVA: Vysokomolekul. Soedin. **2**, 365 (1960).
55. —, S. P. MITSENGENDLER, and V. N. KRASULINA: J. Polymer Sci. **53**, 217 (1961).
56. — —, and K. M. ALEYEV: Vysokomolekul. Soedin. **2**, 1811 (1960).
57. —, and G. V. RAKOVA: Vysokomolekul. Soedin. **3**, 1482 (1961).
58. KROPACHEV, V. A., B. A. DOLGOPLOSK, and N. I. NIKOLAEV: Doklady Akad. Nauk S.S.S.R. **115**, 516 (1957).
59. KROPACHEVA, E. N., B. A. DOLGOPLOSK, and E. M. KUZNETSOVA: Doklady Akad. Nauk S.S.S.R. **130**, 1253 (1960).
60. KUNTZ, I.: J. Polymer. Sci. **54**, 569 (1961).
61. —, and A. GERBER: J. Polymer Sci. **42**, 299 (1960).
62. MARGERISON, D., and J. P. NEWPORT: Trans. Faraday Soc. **59**, 2058 (1963).
63. MATTHEWS, F. E., and E. H. STRANGE: British Patent 24, 790 (1910).
64. MILLER, M. L.: J. Polymer Sci. **56**, 203 (1962).
65. —, and C. E. RAUHUT: J. Polymer Sci. **38**, 63 (1959).
66. MINOUX, J.: Rev. Gen. Caoutchuc. **39**, 779 (1962).
67. MITSENGENDLER, S. P., K. M. ALEEV, L. L. DANTSIG, and A. A. KOROTKOV: Vysokomolekul. Soedin. **5**, 212 (1963).
68. MORITA, H., and A. V. TOBOLSKY: J. Am. Chem. Soc. **79**, 5853 (1957).
69. MORTON, M., E. E. BOSTICK, and R. LIVIGNI: Rubber and Plastic Age **42**, 397 (1961).
70. — —, and R. G. CLARKE: J. Polymer Sci. A **1**, 475 (1963).
71. —, and F. R. ELLS: J. Polymer Sci. **61**, 25 (1962).
72. —, E. E. BOSTICK, R. A. LIVIGNI, and L. J. FETTERS: J. Polymer Sci. A **1**, 1735 (1963).
73. —, L. J. FETTERS, and E. E. BOSTICK: J. Polymer Sci. C **1**, 311 (1963).
74. — — J. Polymer Sci. A **2**, 3311 (1964).
75. —, A. A. REMBAUM, and J. L. HALL: J. Polymer Sci. A **1**, 461 (1963).
76. MULVANEY, J. E., C. G. OVERBERGER, and A. M. SCHILLER: Fortschr. Hochpolymer-Forsch. **3**, 106 (1961).
77. NEPOMNYASCHII, A. I., and KH. S. BAGDASAR'YAN: Kin. i Kataliz. **4**, 198 (1963).
78. O'DRISCOLL, K. F.: J. Polymer Sci. **57**, 721 (1962).
79. —, R. J. BOUDREAU, and A. V. TOBOLSKY: J. Polymer Sci. **31**, 115 (1958).
80. —, and A. V. TOBOLSKY: J. Polymer Sci. **31**, 123 (1958).
81. — — J. Polymer Sci. **35**, 259 (1959).
82. — — J. Polymer Sci. **37**, 363 (1959).

83. Overberger, C. G., E. M. Pearce, and N. Mayes: J. Polymer Sci. 31, 217 (1958).
84. — — — J. Polymer Sci. 34, 109 (1959).
85. —, and A. H. Schiller: J. Polymer Sci. C 1, 325 (1963).
86. Owens, F. H., W. L. Myers, and F. E. Zimmerman: J. Org. Chem. 26, 2288 (1961).
87. Rakova, G. V., and A. A. Korotkov: Doklady Akad. Nauk S.S.S.R. 119, 982 (1958).
88. Rembaum, A., R. C. Morrow, and A. V. Tobolsky: J. Polymer Sci. 61, 155 (1962).
89. Roovers, J., and S. Bywater: Unpublished.
90. Sakurada, Y., et al.: J. Polymer Sci. B 1, 633 (1963).
91. Schlenk, W., J. Appenrodt, A. Michael, and A. Thal: Ber. 47, 473 (1914).
92. —, and J. Holtz: Ber. 50, 262 (1917).
93. Sinn, H., and F. Bandermann: Makromol. Chem. 62, 134 (1963).
94. —, C. Lundborg, and O. T. Onsager: Makromol. Chem. 70, 222 (1964).
95. —, and O. T. Onsager: Makromol. Chem. 55, 167 (1962).
96. —, and F. Patat: Angew. Chem. 75, 805 (1963).
97. Sinn, V., and J. Minoux: Compt. rend. 251, 2020 (1960).
98. Spirin, Yu. L., et al.: J. Polymer Sci. 58, 1181 (1962).
99. —, A. R. Gantmakher, and S. S. Medvedev: Vysokomolekul. Soedin. 1, 1258 (1959).
100. — — — Doklady Akad. Nauk S.S.S.R. 146, 368 (1962).
101. —, D. K. Polyakov, A. R. Gantmakher and S. S. Medvedev: Doklady Akad. Nauk S.S.S.R. 139, 899 (1961).
102. — — — — Vysokomolekul. Soedin. 2, 1082 (1960).
103. Stavely, F. W., et al.: Ind. Eng. Chem. 48, 778 (1956).
104. Stearns, R. S., and L. E. Forman: J. Polym. Sci. 41, 381 (1959).
105. Stroupe, J. D., and R. E. Hughes: J. Am. Chem. Soc. 80, 2341 (1958).
106. Szwarc, M.: J. Polymer Sci. 40, 583 (1959).
107. — Advances in Chem. 34, 96 (1962).
108. — Ber. Bunsen ges. 67, 763 (1963).
109. —, M. Levy, and R. Milkovich: J. Am. Chem. Soc. 78, 2656 (1956).
110. Tobolsky, A. V., and R. J. Boudreau: J. Polymer. Sci. 51, s 53 (1961).
111. —, and D. B. Hartley: J. Polymer Sci. A 1, 15 (1963).
112. —, and C. E. Rogers: J. Polymer Sci. 38, 205 (1959).
113. — — J. Polymer Sci. 40, 73 (1959).
114. Völker, T., A. Neumann, and U. Baumann: Makromol. Chem. 63, 182 (1963).
115. Weiner, M., G. Vogel, and R. West: Inorg. Chem. 1, 654 (1962).
116. Welch, F. J.: J. Am. Chem. Soc. 81, 1345 (1959).
117. Wenger, F.: Chemistry and Industry 1959, 1094.
118. Wiles, D. M., and S. Bywater: Polymer 3, 175 (1962).
119. — — Chemistry and Industry 1963, 1209.
120. — — J. Phys. Chem. 68 1983 (1964).
121. — — Trans. Faraday Soc. (in press).
122. Wittig, G., F. J. Meyer, and G. Lange: Ann. 571, 167 (1951).
123. Worsfold, D. J., and S. Bywater: J. Chem. Soc. 1960, 5234.
124. — — Can. J. Chem. 38, 1891 (1960).
125. — — Makromol. Chem. 65, 245 (1963).
126. — — Can. J. Chem. 42, 2884 (1964).
127. Ziegler, K.: Angew. Chem. 49, 499 (1936).
128. Zutty, N. L., and F. J. Welch: J. Polymer Sci. 43, 445 (1960).

SPRINGER-VERLAG
BERLIN HEIDELBERG GMBH

Beide Teile lieferbar

Biochemisches Taschenbuch

Herausgegeben von Professor Dr.
H. M. Rauen, Münster/Westf.
Mit einem Geleitwort von Prof.
Dr. Richard Kuhn, Heidelberg

Zweite, völlig neubearbeitete
Auflage
Unter redaktioneller Mitarbeit
von Marianne Rauen-Buchka

In zwei Teilen, die
nur zusammen abgegeben werden
I. Teil: Mit 151 Abbildungen
XII, 1060 Seiten 8°. 1964
II. Teil: Mit 166 Abbildungen
VIII, 1084 Seiten 8°. 1964
Ganzleinen DM 156,–

An der bewährten Einrichtung dieses Taschenbuches, das dem Biochemiker rasch unentbehrlich geworden ist, hat sich auch in der zweiten, wesentlich erweiterten Auflage nichts geändert. Unter dem Leitgedanken „Planung – Experiment – Auswertung" sind sämtliche Daten zusammengestellt, die für diese drei Arbeitsstadien als notwendig erkannt wurden. Tabellen, Diagramme, Abbildungen und Texte enthalten die wichtigsten Angaben über die biochemischen Stoffe und Vorgänge. Die zahlreichen Hinweise auf monographische Zusammenfassungen und Originalarbeiten ermöglichen genaue Orientierung über die speziellen Arbeitsgebiete, und ausführliche Darstellungen in den physikalisch-chemischen und mathematisch-statistischen Kapiteln verhelfen dem Biochemiker zur exakten Kritik an seinen Ergebnissen. Das Biochemische Taschenbuch wendet sich jedoch nicht allein an Biochemiker und physiologische Chemiker im engeren Sinne. Es werden auch solche Daten und theoretischen Ausführungen gebracht, die für die Angehörigen der angrenzenden Disziplinen (experimentelle Medizin, Pharmakologie, Bakteriologie, Pharmazie, Zoologie, Botanik) wesentlich sind, d. h. aller der Wissensgebiete, die sich mit den Funktionen der lebenden Zelle beschäftigen.

I. Teil: **Inhaltsübersicht:** International vereinbarte Regelungen für Abkürzungen und Nomenklatur in der Biochemie. – Mathematisch-physikalische Hilfsmittel. – Stoffwerte: Atomgewichte. Haupttabellen biochemischer Verbindungen (Niedermolekulare organische Verbindungen. Makromolekulare Verbindungen). Spezifische Drehung organischer Verbindungen. Absorptionsspektren. Isotopenhaltige organische Verbindungen von biochemischem Interesse. – Sachverzeichnis.

II. Teil: **Inhaltsübersicht:** Räumliche Struktur der Stoffe. – Physikalische Chemie. – Radioaktivität. – Tierversuche. – Körper- und Zellbestandteile. – Biologische Strukturen. – Biologische Funktionen. – Biochemische Arbeitsmethoden. – Statistische Auswertungsmethoden. – Sachverzeichnis.

■ **Bitte Prospekt anfordern!**